DELETE

PAUL ATKINSON

A DESIGN HISTORY OF COMPUTER VAPOURWARE

BLOOMSBURY
LONDON · NEW DELHI · NEW YORK · SYDNEY

Bloomsbury Academic
An imprint of Bloomsbury Publishing Plc

50 Bedford Square	175 Fifth Avenue
London	New York
WC1B 3DP	NY 10010
UK	USA

www.bloomsbury.com
First published 2013
© Paul Atkinson 2013

All rights reserved. No part of this publication may be reproduced or transmitted in any form or by any means, electronic or mechanical, including photocopying, recording, or any information storage or retrieval system, without prior permission in writing from the publishers.

Paul Atkinson has asserted his right under the Copyright, Designs and Patents Act, 1988, to be identified as the author of this work.

No responsibility for loss caused to any individual or organization acting on or refraining from action as a result of the material in this publication can be accepted by Bloomsbury Academic or the author.

British Library Cataloguing-in-Publication Data
A catalogue record for this book is available from the British Library.

ISBN: HB: 978 0 85785 346 2
PB: 978 0 85785 347 9

Library of Congress Cataloging-in-Publication Data
A catalog record for this book is available from the Library of Congress.

Typeset by Apex Covantage, LLC
Printed and bound in China

DEDICATION

To Mum, for her love and support until the very end,
and to my son Isaac, whose light burnt so brightly but for too short a time.

Contents

Introduction	viii
IMAGINED MACHINES	2
MAINFRAMES AND MINICOMPUTERS	
Difference Engine	12
Analytical Engine	20
Hofgaard Machine	26
Nordsieck Computer	30
Saab D2	36
Honeywell Kitchen Computer	44
CTL Modular Three Minicomputer	52
PERSONAL AND PORTABLE COMPUTERS	
IBM SCAMP Design Model	62
IBM Yellow Bird	68
IBM Aquarius	74
Xerox Notetaker	80
IBM 'Atari' PC	86
Sinclair QL+	92
Dragon Professional	100
PEN COMPUTERS	
Xerox Dynabook	108
Apple Figaro	114
Sun Modular Computer	124
GO PenPoint Computer	130
IBM Leapfrog Tablet Computer	136
EO Magni Personal Communicator	142
DualCor cPC	150
MOBILE COMPUTERS	
Sinclair Pandora Laptop	160
Phonebook	168
Siemens PIC	176
Psion Halo and Ace	182
Compaq Dualworlds Notebook	188
Pogo nVoy Communicator	196
Palm Foleo	204
THE AGENCY OF IDEAS	210
Timeline	218
Acknowledgements	224
Picture Credits	226
Notes	228
Selected Bibliography	234
Index	238

Introduction

This is a book about vapourware. This instantly raises two questions—what is vapourware and why write a book about it?

Vapourware is a term that describes future computer products that are announced by a company but which ultimately fail to appear. This might seem a straightforward concept, but in the world of the history of computing it seems nothing is ever straightforward. The arguments over what the term 'vapourware' should encompass rage on and on, as the history tabs on the Wikipedia page for vapourware (or vaporware in the United States) testify. Because of the unresolved nature of its definition, the term's usage remains problematic. So much so that one contact I interviewed refused permission to reproduce images of his work in this book because it had the word 'vapourware' in the title. The ensuing discussion made it clear to me that there is a need to clarify and justify my use of this term.

There are numerous reasons why a particular computer prototype may not have gone into production (as will be shown), but one of the most common applications of the term 'vapourware' is in a situation where a computer manufacturer or software house repeatedly and knowingly announces the imminent launch of a new product despite it being a long way from actual production or even completely non-existent. In other words, one of the main uses of the term is to describe a deliberate attempt to mislead the buying public.

Why would a manufacturer do such a dishonest thing? It may be that the company is aware of a lengthy delay in the development of their new product and is concerned that their existing customers will abandon them and buy from their competitors instead. By announcing a new product, a manufacturer might hope to persuade customers to wait longer than they otherwise would before switching loyalty. On a lesser scale, companies have been known to hint at forthcoming products for no other reason than to keep their name in the limelight and in the forefront of people's minds. In other words, purely for the free advertising it affords—a successful, if misleading, marketing practice. Another possible reason for prematurely announcing products is to attract potential investors or to put pressure on tentative offers of funding from a venture capitalist or business development partner by publicly associating them with the project, making it more difficult for them to later withdraw their support. Other companies have been even more underhanded when, desperate to obtain funds to bankroll initial production runs, they have advertised products as if they were readily available by mail order and banked the money that arrives with orders to fund its development, leaving customers waiting for months for a product that might never see the light of day.

With such negative connotations, it is no wonder many people see vapourware as an undesirable thing with which to be associated. But the failure of an announced product to reach the market is by no means always an intentional act. Rather, it is often the result of a series of unfortunate events. The pace of technological development, especially in a market such as computers, is such that by the time a particular product is ready for market, more than one company has found that the technology used is out of date or even obsolete. In order to meet product launch deadlines, marketing material may have been already printed and even distributed to the industry press in advance of the launch to meet their editorial and publishing deadlines way before the decision not to launch the product is made. Who could blame any company in such a situation for not throwing good money after bad by halting production? In other cases, products have been fully developed and an initial production run made ready for sale when at the final moment, venture capital or internal budgets have been withdrawn for political reasons, leaving thousands of products to be scrapped. Yet others have announced forthcoming products in all good faith and then have been faced with an unforeseen and insurmountable technical problem preventing its launch. If deliberate intention to mislead is a fundamental part of the definition of vapourware, then what are these cases? Those not reaching markets because of technical problems could arguably be labelled 'product failures', but those encountering purely financial difficulties have often created functional products that may well have succeeded had they had the chance to be judged by the buying public. In that case, the term 'failure' seems a little unfair.

Another element of the definition of vapourware is concerned with the public announcement of forthcoming products by their intended manufacturers. But there are cases where an internal product development project was common knowledge throughout a large corporation, as well as to a circle of close customers and business partners and even third-party software developers and technical journalists. Often, working prototypes from these projects were demonstrated to invited audiences at public exhibitions, but no formal public announcement was made before the project was dropped. In these cases the audience expecting the product still numbered in the thousands. So the issue of what constitutes an announcement to an expectant audience for a product is not as clear-cut as it might be.

There are also other future products that have been openly publicized but for which there has never been any intention to go into production and consequently never any intention to mislead. Research-led projects, widely disseminated through academic papers for example, have often been based on completely feasible products just waiting for the existing or emerging technology to reach a suitable state of maturity. Because of being openly discussed, these dreams of future products have often inspired many other products along the way even if they have not eventually reached production themselves. Even though not manufactured, they remain important, influential designs.

So what term can be used to unite and describe all these different cases? Whether they are computer dreams, feasible concepts, prototypes, or failed products, the only thing that ties them all together and acts as a common denominator is that they didn't, in the end, make it to market. In this respect, they are all vapourware—tantalizing glimpses of future products, of what might have been had circumstances been different. The term 'vapourware' should not be limited to cases with negative connotations. The comparable terms 'hardware' and 'software' are not value-laden in this way, be it 'firmware', 'shareware' or 'freeware'. Neither is the slightly ridiculous term 'wetware' (used to describe the human operators and programmers of computer systems). The exceptions are 'spyware' or 'malware', but these are indeed deliberately designed with malicious intent.

I would therefore argue that the term 'vapourware' could and should be used to describe any feasible computer product of which an audience is made aware but which does not then result in an actual product marketed for sale.

So, having defined vapourware in this way, why write a book about it? Well, for one thing, it provides an entertaining view of an alternate world, of products that could well have been available, of designs that nearly and clearly could have been realized were it not for the vicissitudes of fortune. As such, it provides an alternative narrative of the history of computing, quite apart from the established and increasingly well-known stories of important successful computers that changed the industry in one way or another. Little has been written focusing on this aspect, which from the perspective of design history is surprising given the discipline's interest in the design process and the development of new forms. Having said that, design history has changed its focus over the past few decades (especially in the United Kingdom), turning its attention from the production of design towards its consumption. Indeed, my own previous work on the design history of computers has concentrated on their consumption in a social and cultural context rather than their production in a technological context. I have argued that the reasons for a particular form of computer succeeding over another have predominantly come down to that product's reception by users—I have favoured the social construction of technology over technological determinism. That is what makes this topic interesting—none of the featured products in this book have undergone that acid test of public opinion.

The fact that these designs haven't been tested in the marketplace does not, however, mean that they have had no impact. Part of the aim of this book is to highlight this fact—showing the extent to which ideas for products rather than products themselves can have influence far beyond that expected—this is the agency of ideas. It ought to be clarified here that although the definition of vapourware I am employing in this book would encompass the development of software as well as hardware, this particular text concentrates almost solely on hardware—the physical designed forms of computers. This is not in any way to suggest that the development of software has not been of fundamental importance in the history of computing or to our understanding and experience of using computers.

The book could have been structured in any number of ways, but rather than settle on a purely chronological organization, I chose to employ a thematic arrangement. A brief history of imaginary machines is followed by a number of case studies placed into four sections—mainframes and minicomputers, essentially large, business or scientifically oriented machines; personal computers, aimed at individuals; pen computers designed for handwriting recognition; and mobile computers designed primarily for ease of portability—and finally a short concluding chapter. This arrangement was not without its problems, given that a number of the case studies could quite easily be placed into two or even three different sections. The Honeywell Kitchen Computer and the CTL Modular Three were both designed to allow use by an individual rather than numerous users, but primarily they were minicomputers acting as part of a larger system. The IBM SCAMP Design Model and the Xerox Notetaker were designed as portable (if heavy) devices, but more importantly, their creation was executed to prove that complex computer systems could be reduced to stand-alone products owned and operated by an individual. Pen computers were a discrete line of products with interfaces designed to recognize handwritten input, but I have included a later, small mobile computer—the DualCor cPC—in this section because it was based on a tablet version of Windows software with handwriting recognition. The complexities and intricacies of multifunctional computers hamper straightforward classification, and this is further confounded when the functionality of mobile phones is added, as is the case with a number of the cases explored.

Researching the information for a book about computers that never went into production has been particularly difficult. Sources have not been easy to locate. Rumours surface with no supporting evidence, and finding that evidence has often entailed arduous detective work. Some of the hard evidence in the form of actual prototypes has long been consigned to the dustbin of history, leaving photographs as the only record. Finding these images has also been an issue. Often, computer corporations are unable to provide such images from their own archives (if they even have archives), or they have been lost in the mists of an endless round of corporate takeovers and mergers, bankruptcies and buyouts. This is not that surprising—there is little reason for a company to keep images or records of products that they never sold. Other, smaller computer companies, usually started by entrepreneurs, computer scientists or engineers, have embarked on ambitious product development projects that for different reasons came to nought. If a product doesn't succeed, people such as these tend to move quickly on to the next challenge—looking forward rather than back—and care little about keeping accurate records of past projects.

I have found that the people who can be relied upon to keep records of all the projects they have ever worked on are the industrial designers. They inevitably have portfolios of evidence, often containing everything from initial concept sketches to final engineering drawings and almost always with a high-quality, well-lit studio photograph of the final prototype. Without those people, much of the content of this book would not have been available, and I am heavily indebted to them all.

Imagined Machines

The word 'vapourware' suggests something ethereal, ephemeral, otherworldly, a mere suggestion of a product rather than a fixed, readily identifiable artefact. By their nature, vapourware products are not full production items that can be purchased by users, and in that respect they do not exist—they are only imagined. They are not, though, fantastic predictions of future technology but rather statements of a fully intentioned tomorrow. They propose an entirely plausible near future, which nevertheless is yet to happen.

In order to locate vapourware in relation to other types of future predictions and put it into a historical context, it would be valuable to explore a brief, general historiography of imagined machines. Man has imagined machines for millennia. While it would be wonderful to be able to provide a complete history of all these imagined machines, that would be an ambitious undertaking way beyond the scope of this book. The purpose of this chapter, then, is to examine how imagined machines have been considered and presented in the past and compare such imaginings with those of today.

There is no way of knowing, of course, quite where imagined machines first appeared. There are, however, some fairly early descriptions of fantastic technology that were written as far back as the thirteenth century. These were written not as fiction (although they may read as such today) but as serious treatises by learned scholars. The English philosopher and friar Roger Bacon studied at Oxford and later lectured there and at the University of Paris about the works of Aristotle, whose views on science influenced medieval thinking until the advent of Newtonian physics. He wrote widely on the principles of optics, mathematics, astronomy and alchemy and was a respected and well-connected thirteenth-century scholar. Bacon has also been represented as a visionary, predicting the invention of a variety of future technologies, although his writings are written not as imaginings but statements of fact.

In a work of uncertain date but which is widely ascribed to Bacon, translated as *Letter on the Secret Workings of Art and Nature, and on the Vanity of Magic*, there are numerous passages describing machines that would today be called the submarine, the automobile and the airplane. He wrote,

> It is possible to build vessels for navigating without oarsmen so that very big river and maritime boats can travel guided by a solitary helmsman much more swiftly than they would if they were full of men. It's also possible to build wagons which move without horses by means of a miraculous force. And I think that the reaping chariots that the ancients used in battle must have been made like this. It's also possible to construct machines for flight built so that a man in the middle of one can manoeuver it using some kind of device that makes the specially built wings beat the air the way birds do when they fly. And similarly it's also possible to build a small winch capable of raising and lowering infinitely heavy weights … it's also possible to build devices for walking on seas and rivers and for touching their bottoms without taking any risks. And Alexander the great doubtlessly used these instruments to explore the ocean floor as the astronomer Etico narrates … In fact there is no doubt that such instruments had already been built in ancient times and are still being built today, except for the flying machine that neither I nor anyone I know has ever seen. However, I do know a scholar who tried to build this instrument as well. It's possible to build an infinite number of bridges, for example which, can be stretched across rivers without using any kind of pillars or supports, and of unheard of machines and inventions.[1]

The eminence of scholarly sources such as Bacon addressing such fantastic ideas points to the line between fantasy and reality being blurred from the beginning.

Detailed descriptions and designs of early imagined machines began to be more widely disseminated from the fifteenth century onwards when they appeared in print in various *Theatrum Machinarum* (Theatre of Machines). Precursors of technical encyclopedias, these were highly popular and indispensible bound handbooks of technical knowledge, containing etchings of explanatory drawings of all kinds of constructed devices published to celebrate the possibilities of technological progress—a kind of Renaissance 'How It Works' manual. Cross-sectional drawings were used to explain windmills, human- and horse-powered treadmills, water-driven pumps and other mechanical aids. Much of what appeared in such books was entirely practical, but some things were more fanciful.

One particularly popular yet totally impractical topic in these books was the search for perpetual motion, a discussion about an imagined machine that goes back as far as the fifth century[2] and was still the subject of much debate at the end of the fifteenth century when it was 'proved' impossible by none other than Leonardo da Vinci. He drew a weighted wheeled mechanism to show the pointlessness of the quest and compared it to alchemy, writing, 'Oh followers of continuous motion, … You belong to the same fold as those who seek gold.'[3] Nevertheless, the discussion flowed well into the eighteenth century. A variety of particularly detailed water-driven designs further popularizing the idea appeared in *Theatrum Machinarum Novum*, published by George Andreas Bockler in 1662. Bockler was the city engineer for Nuremburg in the mid-seventeenth century and a prolific publisher of well-illustrated manuals. Bockler's illustrations detailed the workings of perpetual motion machines in which overshot waterwheels powered a grindstone as well as an Archimedean screw, lifting the water back up to a reservoir to drive the waterwheel and grindstone again. At that time, with science lacking an understanding of the physical laws about the conservation of energy, there seemed to be no understandable reason why such a device would not work. Hence, Bockler's perpetual motion machines were described as matter-of-factly as windmills and treadmills, and the boundary between fiction and reality remained blurred.

The celebrated Italian Renaissance genius Leonardo da Vinci was perhaps one of the most famous inventors of imagined machines. He was apprenticed in 1466 at the age of fourteen to the artist Verocchio in Florence, where he was likely trained in chemistry, metallurgy, mechanical engineering, architecture and carpentry as well as drawing, painting and sculpture among other skills. He produced thousands of drawings of contraptions of all kinds between the late fifteenth and early sixteenth century, only a quarter of which are believed to remain in existence today. His machines served many purposes, ranging from clock mechanisms, stone-cutting machines, drills and hydraulically operated saws to giant war machines, including tanks and self-loading, rapid-fire crossbows. Some of his most widely reproduced drawings, though, are the ones that have been picked out as prophetic predictions—the fantastic imaginings of self-propelled vehicles prefiguring the automobile; various 'ornitoterri' designed to give man mechanical, jointed wings, driven by pedals, screws and pulleys; fixed-wing gliders; and his 'air screw', a human-powered precursor to the helicopter. It is these 'divinations of future developments' that made Leonardo da Vinci the focus of much analysis in the history of technological ideas.[4]

The rapid expansion of scientific exploration and discovery in the nineteenth century saw major advances in scientific understanding and brought an awareness of technological progress to the forefront of popular culture. The French author Jules Gabriel Verne was a master of extrapolating current technology into the future and is seen by many as the father of science fiction. Verne's writings contain many examples of

Perpetual motion machine from G. A. Bockler's Theatrum Machinarum Novum, *1662.*

Leonardo da Vinci's designs for a military tank and the 'air screw', an imagined helicopter, c. 1490.

imagined machines. In particular, his series of books now referred to as the 'extraordinary voyages'—including *Journey to the Centre of the Earth* (1864), *From the Earth to the Moon* (1865), *Twenty Thousand Leagues Under the Sea* (1870) and *Around the World in Eighty Days* (1872)—predicted space travel, aircraft, skyscrapers, automobiles and even fax machines long before they became a reality.[5] His other machines, such as steam-powered automata or the fantastic vessel Captain Nemo's *Nautilus*, were extrapolated forward from existing technology: Prototype submersibles had been built in the eighteenth century, and it is possible that the *Nautilus* was inspired by the USS *Alligator*, an early American Civil War submarine built in 1861.

The historian of technology George Basalla analysed all kinds of imaginary machines as part of his text *The Evolution of Technology* and classified them into three categories: technological dreams, impossible machines and popular fantasies. In Basalla's view, technological dreams range from unchallenging extrapolations of current technology to far-reaching technological visions; impossible machines are not necessarily based on any existing technology, whereas fantasies appearing in the popular press have become standard fare for science fiction and cartoons as well as science and technical journalism. The value of this classification is not clear, but Basalla's point was that despite their imaginary nature, such machines contribute to a much wider understanding of technology, providing 'a superfluity of novel artefacts from which society makes selections'.[6] As Basalla points out, technologies might be chosen for development irrespective of their current level of possibility or even reality.

In Basalla's terms, then, imagined machines such as vapourware would be examples of a technological dream, a low-level extrapolation of existing technology; but such a label is misleading. There is a potential distance between possibly unattainable dreams and entirely feasible products that is, for me, too far. As will hopefully become clear through the course of the content of this book, some examples of vapourware are more imaginary than others. Some are indeed a complete fiction and exist only in the minds of their inventor and later the unsuspecting public at the receiving end of deliberately misleading announcements. Yet other examples actually do exist—admittedly sometimes only in the form of proof of concept prototypes, but at other times as fully functioning initial batches of finished products that for various reasons are withheld from the customer. Even here, they are still imagined as machines that would be bought and used.

So perhaps with respect to vapourware, a key question is where is the boundary between fiction and fact, between extrapolation and fantasy?

Basalla is not alone in realizing the powerful potential of predictions. As has been noted with reference to science fiction, futuristic predictions 'can hold a mirror up to ourselves and society and show how we react. Science Fiction is that speculation about the impact of science, technology and socio-political change on us, hence the alternative phrase "Speculative Fiction."'[7] And in the preface to *Yesterday's Tomorrows*, the authors point out that 'visions of the future are not passive reflectors of historical reality. They can also actively encourage or perpetuate certain attitudes or models of behavior.'[8] It is not only science fiction that has done this, but design too.

There was a time when predicting the future seemed a more serious business than it does today. In the early part of the twentieth century, the certainties offered by modernism and its blind faith in the potential of progressive technology to serve mankind pointed to a mechanically aided future where we would all benefit. But it was not long before that future was posited on an expanding capitalism. A plethora of public events and promotional films appeared from the late 1930s onward, promising people that buying into corporate consumerism would be a panacea. The 1939 New York World's Fair was a wonderful example, containing one of the most ambitious exhibits of all—'Highways and Horizons'. Designed by Norman Bel Geddes on behalf of General Motors (GM), the exhibit consisted of four interconnected buildings containing numerous displays of the latest GM cars (including a transparent acetate car) and a range of household goods. The main attraction, though, and the one that people queued for hours to see, was Futurama—a fairground ride of 600 chairs, which moved visitors over a scale model of future cities of 1960 connected by 100-mph seven-lane freeways.[9] The enormous model, which was the hit of the fair, covered 35,000 square feet and was a third of a mile long. 'Geddes' Futurama took you on an imaginary flight across America from coast to coast … There were frequent changes of scale, to give you the sensation of swooping down closer. The last you saw of the diorama was a particular street intersection of the future, up close … and there you were at that precise intersection, life-sized.'[10] Through the clever use of scale, Futurama provided an 'immersive experience' that blurred fantasy and reality. As design historian Nic Maffei put it, 'Its key design aim was to gently and incrementally transport the viewer into an imagined future, albeit

The Futurama ride, part of the General Motors 'Highways and Horizons' exhibit at the World's Fair in New York, 1939.

The Frigidaire 'Kitchen of the Future', 1956

one which was dominated by the mass consumption of General Motors' automobiles and an unyielding ideology of technological progress.'[11]

The future became a staple part of popular culture, full of imagined machines that would make life so much easier. The Daily Mail Ideal Homes Exhibition's 'House of the Future' in 1956 predicted (quite accurately) that by 1980, almost everything in the house could be plastic and all electronic equipment would be operated by remote control. In the same year, General Motors' *Design for Dreaming* film featured Frigidaire's 'Kitchen of the Future': punch-card-controlled computers that dispensed and mixed ingredients, refrigerators with motorized revolving shelves and ovens with glass domes. The American kitchen, as Dag Spicer noted, is 'a particular cultural "meme" that seems to never die … a site where technology, in the name of "increased leisure time," bumps its head against the real world of driving kids to school, making dinner, and washing dishes'.[12] Philco-Ford's 1967 film *1999 A.D.* looked forward to meals planned by a computer, ordered by push button and cooked within seconds, perfect for 'a society rich in leisure and taken for granted comforts'.[13]

The other area, particularly of American life, affected by technological visions of the future revolved around the 'car of tomorrow', long noted as 'an essential prop to a society built around the private automobile'.[14] As a part of the wider canon of futuristic transport including huge monorail systems and rocket trains, convertible jet helicopters and ocean-crossing superliners portrayed on the covers of magazines such as *Science and Invention* and *Modern Mechanix*, the future car was presented as the constant companion of the aspirational consumer. With the appearance of 'automotive stylists' in the 1930s, the futuristic concept car moved from the pages of popular science magazines into the designer's studio. Models and prototypes of streamlined vehicles from Bel Geddes and Buckminster Fuller inspired the form of the car of tomorrow for decades. As the phenomenon reached its peak in the 1950s and 1960s, the 'visioneers' designing these 'dream cars' were encouraged 'to look further into the future than ever before',[15] and their designs were promoted far more widely. The celebrated stylist Harley Earl even assembled 'Motorama', a travelling auto extravaganza, which from 1952 took General Motors' latest futuristic concepts around the country. While such events can be dismissed as purely publicity stunts to increase interest and hence sales, they also served a very pragmatic purpose. Whenever a manufacturer launched a car that was 'ahead of time', there was a very real

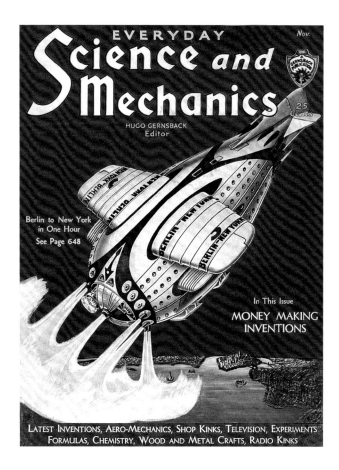

Future transport predictions, Everyday Science and Mechanics, *1931, 1932.*

General Motors GM–X Stiletto concept car, 1964.

danger that public reaction would be less than favourable, 'but if the new idea was displayed as a lavish and unattainable dream car, it became, by the time it went into production not just acceptable, but highly desirable.'[16] This approach gave the consumer the best of both worlds. An aspirational future full of hope and promise, as well as (when watered down) a dream-fulfilling vehicle for today: the fantasy made real.

In the introduction to a recent publication accompanying a major exhibition of science fiction, Mike Ashley wrote,

> We live in a science-fiction world. We take for granted items that were the dreams of a few visionaries only a generation or two ago—personal computers, organ transplants, mobile phones, space travel, virtual reality—all of which have enhanced and improved our lives.[17]

Of course, we don't really live in a science fiction world; we live in a very real world. But the point is that the boundaries between the worlds of science fiction and science fact are in no way fixed but entirely fluid. Much of what was previously considered to be outlandish fantasy is now reality and some of what remains fantasy now might move into the realm of the real at any point. This happens on a regular basis and is something we accept without question. The boundaries between fact and fiction change over time as the market and technology progresses. Predictions that were merely speculation or extrapolation are routinely proved true.

Witness, for example, the rather incredible case of the videophone. The concept of face-to-face communications over great distances goes back a remarkably long way. The well-publicized achievements of the prolific inventors Thomas Edison and Nikola Tesla in the closing years of the nineteenth century sparked an electrical and radio-driven version of the space race—to the extent that they even appeared as leading characters in science fiction stories.[18] Media interest in electrical technology was such that conjecture on its possibilities soon became commonplace. In 1879, shortly after the invention of the telephone, a cartoon featuring an imagined 'Edison's Telephonoscope' that transmitted images along with sound appeared in *Punch Almanak*. An early science fiction story of 1911, *Ralph 124C 41+* by Hugo Gernsback, who later went on to found the hugely successful and much copied magazine *Science Wonder Stories*, featured the 'Telephot', a proto-videophone connecting distant planets.

The idea reached a wider audience in 1914 when Victor Appleton wrote *Tom Swift and His Photo Telephone*, but still the videophone remained strictly fantasy for a great many years. Regular appearances in *Flash Gordon* from the 1930s, *Dan Dare* in the 1950s and *The Jetsons* in the early 1960s continued to present the idea as pure fiction, although the technologies to enable such a device—mechanical television coupled with telegraphy, radio or telephony—were available from the late 1920s. The German post office trialled a public videophone service in the late 1930s, but it was dropped when the Second World War took over events. AT&T's Bell Laboratories had also been working on similar devices since the early 1930s and eventually produced a real working videophone system called 'Picturephone' that was demonstrated at the World's Fair in New York in 1964, linking the East and West Coasts of America. A futuristic version of the Bell Picturephone even made an appearance in Stanley Kubrick's *2001: A Space Odyssey*, but trials of a real (and reportedly expensive) Picturephone system linking New York, Washington DC and Chicago failed to ignite public interest and the system was quietly closed down.[19]

So public interest and the constant representation of the videophone in mainstream fiction might lead one to believe that a high demand for such a product was present and that the opportunity to use such a system would be welcomed with

The AT&T Picturephone demonstrated at the World's Fair, 1964.

open arms, but the truth of the matter is that despite the current availability of relatively free internet-based alternatives such as Skype and FaceTime, the use of video-based communications remains, at best, limited. Historians will no doubt continue to argue about the extent to which the forces of technological determinism or the social construction of technology have been most in play in the case of the videophone's development and failure to become adopted on a wide scale, but the relevance of the story is that at different points in time, the videophone went from being a futuristic piece of science fiction, to being a feasible prediction based on the extrapolation of existing technologies, to being an experimental prototype and finally a real product. It went from being a purely imaginary machine to a real machine, and yet the boundaries between the two are not quite as clear as might be imagined.

Quite why some technologies make the transition from fantasy to reality and others never really do is an interesting question and one which historians of technology have struggled to answer. The issue is of some importance, as the old notion that the technologies which are ultimately successful achieve that success because they are without question 'the best' is rightly considered unacceptable. As Hans-Joachim Braun said, 'Failed innovations are just as important, and possibly even more so than, successful ones'[20] because they provide a far more realistic view of how a particular technology developed, especially given that failure can be for technical, economic, social, political or cultural reasons.

Various explanatory models to explain technological failures have been put forward. Trevor Pinch and Wiebe Bijker's theories of the social construction of technology argue that the success of any technology is subject to that technology being perceived as the solution to a problem by the relevant group of users it aims to serve. The problem they identify for manufacturers trying to take this fact into account is one they term 'interpretive flexibility'.[21] Essentially, this is the view that any product can mean different things to different people in different circumstances. In other words, it doesn't matter how well a technological product actually works if it doesn't 'look right' or 'feel right' to the customer or match with their different expectations. In effect, the technology doesn't work *for them*. Not everybody is convinced by these explanations, though. Kenneth Lipartito's concern is that failure is often seen as a prelude for improvement, with the failures themselves 'relegated to their proper place, either the dustbin of history or the lesson book of progress'.[22] The problem he sees is that when one assumes failure is self-evident or a part of an evolutionary process, one all too easily replaces the unacceptable assumptions of technological determinism with an unwitting social determinism, which denies 'the deeper cultural paradigms that shape technological change'.[23]

An alternative view of the causes of technological failure is the theory of 'path dependency'. In contrast to both social construction and technological determinism, path dependency accepts that failure can be dependent on a whole sequence of incidents in which 'important influences upon the eventual outcome can be exerted by temporally remote events, including happenings dominated by chance elements rather than systematic forces'.[24] An example of path dependency can be seen in the case of the QWERTY keyboard. Paul A. David suggests that despite being proven less effective than a number of alternatives that appeared at a very similar time, the QWERTY keyboard became accepted as the dominant arrangement because of an emerging infrastructure of production that supported it. Companies quickly emerged to train operators to touch-type using the QWERTY system, and once in place, it proved an impossible job to undo the advantage the system had gained. There was little point in manufacturers adopting a different keyboard when the vast majority of typists were inculcated into the QWERTY system and the majority of offices had invested in QWERTY typewriters. Consequently, the initial selection of one keyboard layout over another can be seen as a 'historical accident'—a 'particular sequencing of choices made close to the beginning of the process. It is there that essentially random, transient factors are most likely to exert great leverage.'[25]

The vicissitudes of technological development, however, are such that even high path dependency can be broken. It seemed highly unlikely to many people, given the long-established, global infrastructure that was in place, that the 35mm film format would be changed at all, let alone disappear as quickly as it did. Early digital images were so poor in quality that 'serious' photographers derided them mercilessly, yet the advantages of instant results and essentially infinite photographs was so attractive to the majority of users that it was readily accepted despite its obvious drawbacks. With all research and development resources diverted from traditional wet film going into improving the resolution of digital cameras, it was not long before the system was perfectly acceptable to professional photographers.

Sometimes, the desire for a particular form of product, even if it is impossible at the time, is a strong enough force for change. Take the case of the cathode ray tube (CRT)—again, a well-established technology firmly in place across the world when the futuristic fantasies of science fiction artists and designers depicted flat-screen televisions in future homes that were so thin as to be nearly invisible and that filled whole walls.[26] Driven by consumer desire to produce ever flatter and larger televisions, a huge amount of research and development money was poured into trying to produce a thinner cathode ray tube. One solution, the Panasonic Flat Vision TV of 1994, a 14-in. cathode ray tube television less than 4 inches deep, split the screen up into 10,000 sections and gave each section its own electron beam source, 'basically replacing one big CRT with lots of tiny ones'.[27] The downsides were that it cost ten times as much and weighed half as much again as a standard 14-in. TV and had 'ravenous power consumption'.[28] Perhaps not surprisingly, it was not a bestseller. The alternative to thin CRTs was of course LCD (liquid crystal display) screens. Although the picture quality was nowhere near that of high-resolution CRTs, once research monies were diverted from developing CRTs and focused on developing LCD technology, acceptable-quality LCD televisions finally reached the market, and it was not long before the resolution was as good and then better than that of CRTs. Once again, the futuristic vision had become reality.

These examples show the value of futuristic visions in driving technological change. The problem is that yesterday's tomorrows look so much more exciting than today's. We imagined going into space, and we ended up walking on the moon. So we imagined going to Mars and Jupiter. There seemed to be no boundaries. The future was presented as holding incredible opportunities. But such forays into the future appear to be anathema today. It is as if we have finally given up, jaded perhaps from years of experience of technology often not turning out as we expected, as if the constant mocking of the naivety of some past forecasts has made us think twice about proposing anything that might happen tomorrow. Perhaps, at the start of the twenty-first century, the reality of everyday life is so far removed from where we thought it might be that there seems little point in second-guessing what the future may hold. Or is it because the technology that has appeared has changed our lives so fundamentally in ways we never imagined that we no longer deign to imagine the future? Is the potential impact of nanotechnology, genetic engineering and biological warfare so frightening that we prefer not to speculate? The reasons are not clear, but we certainly do not seem to embrace the future with the same optimism we used to. Whatever happened to those big ideas? Menial tasks done by uncomplaining robotic servants; worrying about what to do with all our spare time when there was no need to work; driverless automobiles and levitating trains on lines spanning oceans; living in domed cities under the sea or on the moon and holidaying on space stations or on Mars, all notions presented on a regular basis throughout the first half of the twentieth century but which now seem conspicuous by their absence. Arthur C. Clarke was right: 'The future isn't what it used to be.'[29]

The *Steampunk Manifesto* laments the lack of such exciting, imaginative futures:

> Today, the only future we are promised is the one in development in the corporate R&D labs of the world. We are shown glimpses of the next generation of cell phones, laptops or MP3 players. Magazines that used to attempt to show us how we would be living in fifty or one hundred years, now only speculate over the new surround-sound standard for your home theater or whether next year's luxury sedan will have Bluetooth as standard equipment.[30]

To a large extent, this description fits vapourware products quite closely. Even intentional vapourware doesn't present itself as an implausible technology from the far future. Who would believe that? Vapourware by its very nature has to be completely believable, given that in many cases what is being described is not only feasible but also extant. This book is not about those fantastic and futuristic visions, but it is about imagined machines. Where those past predictions looked far ahead, vapourware looks instead to a very near future—days, weeks, months, a few years at most. Nevertheless, the study of vapourware still reveals a world that could have been, a world of alternative possibilities, even if it isn't that far removed from our own. Vapourware still represents an imagined future, even if it is one that is just around the corner. A history of vapourware is a history of intention. As such, vapourware has a completely valid place in the pantheon of imagined machines.

Mainframes and Minicomputers

The driving force behind the mechanization of mathematical calculation and information processing did not come from a single source but emerged from a number of different directions, ranging from the purely scientific quest for knowledge to the military desire for power and the requirements of business for accounting, record keeping and organization. Hand-operated, small-scale desktop mechanical calculators were used successfully for many years for a wide range of mathematical and accounting purposes, and early data processing equipment such as the Hollerith tabulating machine enabled the results of the American census of 1890 to be available far more quickly than in the previous census and at far less cost.[1] But the continuous processes of industrial expansion and globalization meant that the requirements of the military, science and business outpaced the capability of such machines to provide the necessary answers. This problem was solved in part by the addition of electrical switching through relays to mechanical calculators. This provided large-scale electromechanical computers used for a variety of military, scientific and business purposes. Alan Turing's 'Bombe' was an electromechanical machine 2.1 m (7 ft) wide, 1.98 m (6 ft 6 in.) tall and 0.6 m (2 ft) deep, weighing around a ton, over 200 of which were used to break the German Enigma code in 1940. During the war, 200 Bombes were constructed to decode German messages.[2] The largest electromechanical computer ever built, IBM's ASCC or 'Harvard Mark 1' of 1944, was 16 m (51 ft) long and 2.4 m (8 ft) high, contained 800 km (500 miles) of wiring and 3,500 relays, and was used to provide ballistics calculations for the US Navy for fifteen years.[3] Other uses of such machines included calculating the positions of stars and planets, simulating aircraft stability, organizing national train timetables, and analysing payrolls for corporate employees, ad infinitum. As time went on, the list of computerized functions grew exponentially. The machines that provided the solutions were hugely expensive, though, and only the largest corporations and national institutions could afford to develop and build them, with time on the computers being sold as a service to outside bodies. Although many companies used them, computers did not become commercial products until later.

This growth in computer usage was accompanied and promoted by a series of technological developments. Electromechanical machines were superseded by much faster electronic computers based on vacuum-tube or valve technology that first emerged from wartime developments. In the United Kingdom, the Colossus was developed at Bletchley Park to break the German Lorenz cipher used for high-level military communications, and in the USA, the ENIAC was developed at the University of Pennsylvania to speed up ballistic trajectory calculations. Indirectly, these developments led to the first commercial electronic computers for business use produced in 1951—the British Ferranti Mk 1 and the American UNIVAC among them. Such valve-based computers started a rapidly expanding industry and they began to be used more and more widely within science, industry and business until the arrival of the first transistor-based machines. Invented in 1947, transistors first started to appear in commercially available computers in the mid 1950s and affected the technical and commercial growth of electronics more quickly than was expected.[4] Their smaller size enabled powerful, smaller computer systems to proliferate throughout offices until transistors in turn were replaced in the 1960s with the integrated circuit and finally the microprocessor, followed by the debut of computers that were cheap enough and small enough to be used as desktop machines.

Associated with the decreasing size and cost of computers was a change in the roles they performed. Large mainframe computers housed in special departments within companies became centralized machines that were accessed by numerous users through remote computer terminals, and eventually, the smaller computer systems enabled by transistors began to appear throughout the office landscape until microprocessor-based computers began to be used by individuals at their own desks. In the process, computers changed from being used purely for corporate-level organization, through managerial departmental control, to individual work production.

Understandably, given the scope and spread of these technological and contextual changes in computing, numerous dead ends were explored and machines developed that did not see the light of day. New, untested technologies were tried that brought their own problems, leading to functional failures or unreliable products that could not be marketed. Existing technologies were often pushed way beyond their time in the face of uncertainty regarding the acceptance of new inventions. For example, some companies continued to try to develop electromechanical relay-based machines after the introduction of valve-based computers because of the lower cost and proven reliability of relays over that of valves. Similarly, when transistors were first introduced, their high price and scarcity meant some companies chose to continue research and development work with valves. Often, before such machines were ready for market, the new technologies had either reduced in cost or become more reliable and readily available, rendering the earlier technology obsolete. Additionally, and quite remarkably, up to the mid 1960s, there was no such thing as industry-wide computer compatibility. Until IBM announced their System/360 in 1964,[5] every manufacturer produced machines that would not work with those from their competitors (and in many cases, not even with other machines they had manufactured themselves). Unsure as to the impact of this announcement, many companies continued along the lines of developing their own unique systems only to find that before these machines could be sold commercially, they were targeting a market that had changed forever in favour of IBM's standardized architecture.

To some extent, mainframe computers and minicomputers are now a thing of the past, their functions replaced by those of much smaller, cheaper, more powerful and widely available machines, although the term 'mainframe' is now frequently applied to servers—a networked computer providing a service to clients. In this context the term 'mainframe' refers to any commercial computer that can support thousands of applications and input/output devices to simultaneously serve thousands of users.[6] The current state of affairs may appear on the surface to have reached a reassuring level of stability, but the vicissitudes of the computer industry have, from the very beginning, been such that nothing could ever be taken for granted, and consequently this industry has been fertile ground for vapourware.

Project **THE DIFFERENCE ENGINE**
Designer **CHARLES BABBAGE**
Client **THE BRITISH GOVERNMENT, UK**
Date of design work **1821–1849**

Construction of the Difference Engine No. 2 built by the Science Museum in London, completed in 2002.

DESCRIPTION **A mechanical calculating machine driven by hand crank, in total consisting of 8,000 precision-machined steel and bronze components in a cast-iron frame, weighing between 4 and 5 metric tons and measuring 3.33 m (11 ft) long by 2.25 m (7 ft) high and 1.25 m (4 ft) deep. On the right-hand side is the control mechanism and crank, the central part with eight vertical columns is the calculating section and the printing section is on the left.**

As the leading expert on the subject, Doron Swade, once wrote, the celebrated mathematician Charles Babbage is equally famous for two things: for inventing vast computers and for failing to build them.[7]

The familiar and often retold version of events is that Babbage and his friend the mathematician and astronomer John Herschel were checking manually prepared mathematical tables in 1821 when the frequently occurring discrepancies caused Babbage to proclaim, 'I wish to God these calculations had been executed by steam.' Realizing the possibilities of the statement, he devoted the rest of his life to creating a machine that could automatically and accurately solve complex mathematical calculations.

Early books of logarithmic tables, such as this example from 1670, were riddled with errors that regularly caused fatal shipwrecks.

The pre-eminent narrative of Babbage's endeavours over the course of his life is that he devoted himself primarily to solving one particular problem—producing reliably accurate calculations for use in logarithmic tables. In fact he had far greater expectations for his calculating engines. Babbage's understanding of the potential impact of such devices went far beyond the automatic calculation of error-free mathematical tables; he realized that a machine could also ensure no errors occurred in transcription by automatically typesetting the results using movable type, proofreading them and then directly printing the results out by impressing the type into blocks of plaster.[8] Furthermore, Babbage correctly envisaged that calculating engines would lead to a new branch of mathematics now known as computational analysis and to some extent presaged Alan Turing's 1936 formalization of the problems of 'computability'.[9] He even developed a unique 'Mechanical Notation' to describe the operation and logic of the complex mechanisms, which was 'a serious attempt at formal symbolic representation of computational logic'.[10]

The reason for the narrative focus on overcoming errors in printed tables was that it provided a very convenient and readily understood account of the problem-solving abilities of the Difference Engine that could be applied to a real and pressing problem: errors in printed tables were not only inconvenient but potentially fatal in the case of navigational charts and were believed to be the cause of many shipwrecks. Such an argument more easily persuaded those approached to fund the construction of such a machine.

In one respect, this approach was highly successful. Design work started in 1822, and with the aid of only a small experimental model, Babbage managed to secure significant amounts of funding from the British government to construct his Difference Engine. On the other hand, limiting the description of the Difference Engine's potential to one of pedestrian utility meant other, highly influential people—among them George Biddell Airy, the Astronomer Royal and consultant to the government on such technologies—declared the device to be 'useless'.[11] The utility argument proposed an improved way of doing something that could already be done, whereas the creation of a new branch of mathematics could achieve something never done before.[12]

Babbage's highly ambitious design for the Difference Engine No. 1 called for an incredible 25,000 precision-engineered parts to be accurately assembled. After a decade of design and development and the spending of the then enormous sum of £17,470 of public money,[13] all that he managed to produce was a prototype mechanism comprising one-seventh of the calculating section. After Babbage argued with John Clements, the engineer who built it, the construction venture collapsed in 1833. Following advice from George Biddell Airy, the government finally withdrew from funding the project in 1842.[14]

At that point, the prototype section was consigned to a glass case in a museum, yet it is a testament to how advanced and influential this design was that forty years after its conception and thirty years after its construction, it was displayed in London alongside other commercially available calculators at the International Exhibition of 1862. The exhibition aimed to showcase the latest advances in technology, and the jurors of the exhibition stated that Babbage's machine was still of 'a higher order' than those available.[15]

Following the collapse of the construction of Difference Engine No. 1, Babbage began in 1834 to work on a far more ambitious project, the Analytical Engine (see Analytical Engine). This device was intended to perform far more complex functions and required Babbage to develop mechanisms which could automatically multiply and divide and be regulated via a convoluted control system. In solving these problems, Babbage was inspired to create a superior design for a Difference Engine.[16]

The plans for Difference Engine No. 2, which was simpler in construction (consisting of only 8,000 parts), were drawn up between 1846 and 1848,[17] and yet the more elegant design could calculate numbers to almost twice as many decimal places (thirty-one versus the sixteen decimal places of Difference Engine No. 1).[18] Babbage offered his plans for the Difference Engine to the government in 1852, but they showed no interest in supporting its development.

The story of the Difference Engine might have ended there were it not for the endeavours of a dedicated team at the Science Museum in London. The team, led by Doron Swade, completed a fully functioning construction of the calculating section of the machine (which alone consisted of 4,000 separate components) that was completed in time for an exhibition in 1991 celebrating the bicentenary of Babbage's birth.[19] Swade wrote that carrying out the ambitious rebuild, an amazing achievement by any standard, 'gave us tremendous respect for Babbage's ability to visualize the operation of complex mechanisms without the aid of

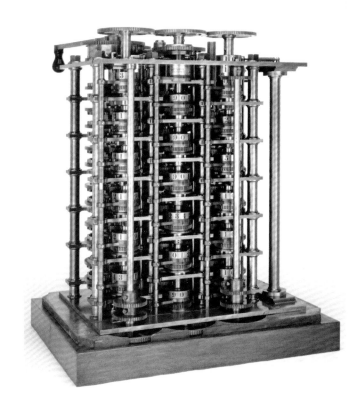

Prototype of part of Babbage's Difference Engine No. 1, built by John Clements between 1824 and 1832.

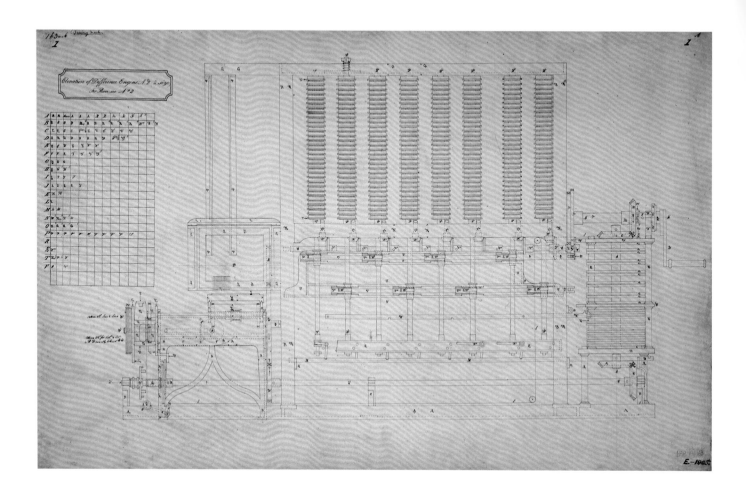

Drawing by Charles Babbage of the side view of Difference Engine No. 2, late 1840s.

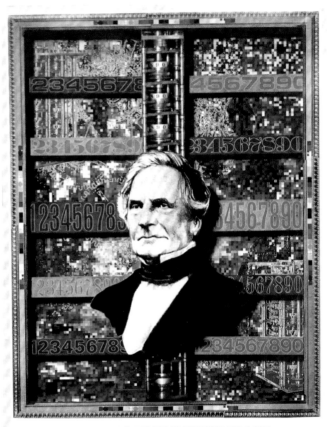

Poster for 'Making the Difference: Charles Babbage and the Birth of the Computer' exhibition, Science Museum, London, 1991.

physical models'.[20] With the use of modern production techniques, but carefully restricted to manufacturing tolerances achievable by nineteenth-century engineers, the result silenced critics who believed that the Difference Engine would never have worked and was incapable of being built. The output apparatus, used for printing and stereotyping the results of calculations, was an integral part of the design and conception of the Difference Engine and consisted of a further 4,000 components. The same team completed this final part of the machine in 2002, and the whole assembly is still on display at the Science Museum today.

Although the Difference Engine remains by strict definition a calculator rather than a programmable computer, Babbage's earliest work on the device evidently marks the start of the age of automatic computation.

Project **THE ANALYTICAL ENGINE**
Client **NONE**
Designer **CHARLES BABBAGE**
Date of design work **1834–1871**

Prototype of part of the Analytical Engine, 1834–1871.

DESCRIPTION **Babbage's Analytical Engine was similar in outward appearance to his Difference Engine in that it was to be an assembly of precision-machined steel and bronze components with number wheels cast in Britannia metal, all housed in a cast-iron frame. But this engine dwarfed his previous designs. The 1840 version of the design would have measured 4.5 m (15 ft) high by 2.5 m (8 ft) deep with a length dependent on the amount of memory. An 'entry level' version with one hundred registers would measure 9 m (30 ft) long—about the same as a medium-sized locomotive.[21] And, like a locomotive, it would have to have been driven by a steam engine.**

Charles Babbage's first design for a calculating engine, although in itself a hugely ambitious and progressive undertaking, was nevertheless a special-purpose machine limited to performing a particular function—the calculation of logarithmic and trigonometric functions. During its extensive development, however, Babbage came to realize the full potential of automatic mechanical computation. Disheartened after his initial attempts to produce a fully functional Difference Engine came to nothing, Babbage turned his attention in 1834 to an even more audacious and wildly ambitious project—the Analytical Engine. With no support from the government, Babbage used his own inheritance to finance the project. It was an endeavour he continued until his death in 1871.

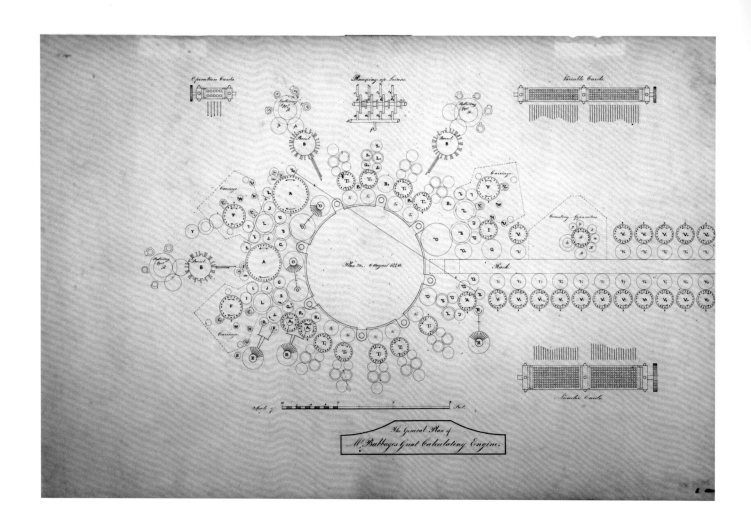

Plan of the analytical engine drawn by Charles Babbage, c. 1840.

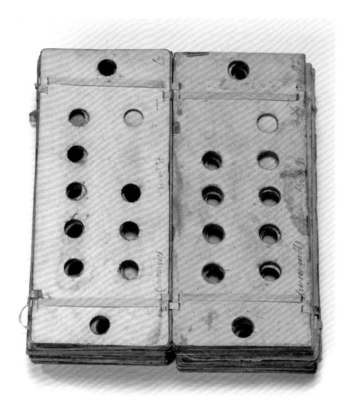
Punched cards for programming Babbage's Analytical Engine, 1834–1871.

Detail of the print mechanism of Babbage's Analytical Engine, 1834–1871.

The design of the Analytical Engine marks the transition from the simple calculation of the Difference Engine to full general-purpose computation (and supports the argument that it was the potential of calculating engines that drove Babbage rather than the preparation of mathematical tables).[22] The Analytical Engine was surprisingly similar in function to a modern general-purpose programmable computer, although like the Difference Engine it used decimal rather than binary notation.

Programmes for the engine to execute were to be entered using a punched card system, in the same way that patterns were created in textiles on a jacquard loom (which later became the standard system adopted by electronic computers using cards developed from Hollerith tabulating machines). Where the Difference Engine was limited to the mathematical process of addition, the Analytical Engine was capable of all four arithmetic functions and, in operation, employed similar processes to modern computers, including serial operation and parallel processing. Borrowing terminology from the textiles industry, the mathematical functions were to be carried out in a separate 'mill', which equates to the central processing unit of an electronic computer, and the results of those calculations were to be held in a 'store', in the same way as a modern computer holds data in its memory. This architecture is essentially the same as the 'von Neumann' architecture that forms the basis of almost every modern computer and was described by John von Neumann in 1945.[23]

The IBM ASCC (Automatic Sequence Controlled Calculator), 1944.

The results produced by the Analytical Engine were to be printed out in hard copy—as a series of punched cards or plotted graphs or, as in the Difference Engine, by automatic typesetting and stereotyping. Compared to the Difference Engine, though, its sheer size meant that there was no way that it could have been cranked over by hand, and the Analytical Engine would have had to have been operated through the mechanical power of a steam engine.

Babbage's influential work on the Difference Engine was a widely known venture in his day and symbolized the major contemporary attempt at automatic computation. After Babbage, it was commonly accepted that mechanical computation was the way forward. By comparison, little was known about his work on the Analytical Engine outside of a narrow circle of his friends,[24] although enough was known that the report by the jurors of the

International Exhibition of 1862 described the machine under development as being of 'a still higher order than the Difference Engine'.[25] It is not surprising, therefore, that after his death the Analytical Engine fell into obscurity and failed to have the impact on technological progress that it so clearly deserved.

That obscurity meant that there is no continuous or direct line of development between the work of Charles Babbage and computers of today, although much has been made of the influence of his work. There is a well-known inspirational tale about how a remnant of the Analytical Engine inspired the development of a highly successful and influential electromechanical calculating machine a century after Babbage's exploration. The story goes that during the Second World War, a postgraduate physics student at Harvard University named Howard Aiken proposed the building of a giant calculating machine to his physics department management. The chairman reported that a lab technician had told him that there was already such a contraption stored somewhere up in the attic. Aiken searched for and finally came across the curious relic—a test piece of Babbage's machine consisting of a number of brass wheels mounted on a mahogany base—and instantly realized that he and Babbage had the same mechanism in mind. However, where Babbage had been starved of funds, Aiken's access to almost unlimited financial resources courtesy of IBM allowed him to accomplish the development of the machine. The result, the IBM ASCC (Automatic Sequence Controlled Calculator), was completed in 1944 and became more commonly known as the Harvard Mark 1. It was a complex machine, containing over 805 km (500 miles) of wiring and 765,000 components. Its programs were stored on long loops of punched paper tape, and it reliably performed ballistics calculations for the US Navy day and night for over fifteen years.

Demonstration model of Difference Engine No. 1 assembled from leftover components by Babbage's son Henry Prevost Babbage, c. 1880.

As beguiling as this story is, the truth is a little less romantic. It is indeed true that there was a section of Babbage's machine at Harvard, but it was actually a demonstration model showing the operating principle of the Difference Engine No. 1, one of a number of pieces assembled from leftover components by his son Major General Henry Prevost Babbage[26] and donated to various universities in the United Kingdom and United States in the 1880s in order to promote his late father's work. And even though Aiken is often referred to as one who was 'directly inspired by Babbage's vision' and went as far as 'presenting himself as one who finally had brought Babbage's dreams to reality',[27] scholarly research has shown that Aiken didn't actually come across Babbage's work until after his own draft proposals for a calculating machine had been drawn up and submitted to Harvard. In fact, despite some similarities in the physical arrangement of components, the Mark 1 'shows precious little influence of Babbage's architecture'[28] and even 'suffered a serious limitation which might have been avoided if Aiken had known Babbage's work more thoroughly'.[29]

Despite the lack of a direct line of development, Babbage's work on the Analytical Engine remains the earliest embodiment of the principles of general-purpose computation and is the compelling evidence on which his appointed status as the 'father of computing' stands.

Project **THE HOFGAARD MACHINE** Designer **ROLF HOFGAARD**
Client **ELEKTRISITETSFIRMAET SØNNICHSEN & CO.** Date of design work **1939–1960**

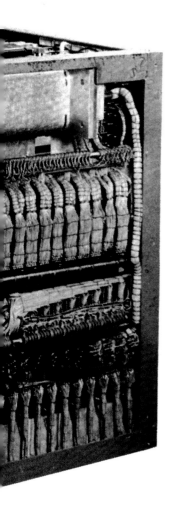

Prototype of the Hofgaard Machine with panels removed, c. 1955.

DESCRIPTION **A desk-sized electromechanical relay calculator with steel frame and panels. Operated via a paper punch tape produced by an adapted electric typewriter.**

Just before the start of the Second World War, the Norwegian engineer Rolf Hofgaard approached the manufacturer of electrical appliances Elektrisitetsfirmaet Sønnichsen & Co. (Electricity Company Sønnichsen & Co.) with a proposition for building a computer that would handle all the accounting for the company. The design, it was thought, would form the basis of a business computer that could be serially produced and sold.

Never having manufactured such a machine before and having relatively limited resources, the company spent the best part of twenty years in a convoluted development program trying to resolve the various technological issues themselves, shunning the interests of experienced computer manufacturers such as IBM, who offered to start a cooperative venture to produce the machine. Eventually and against the odds, they succeeded and produced a viable working prototype that proved to perform exceedingly well. But by that time, it was too late. Relay-based computers were now old technology and the world had moved on.

A Francis Sønnichsen brochure for lead-acid batteries, c. 1950s.

Francis Sønnichsen, a young and well-educated Norwegian entrepreneur, founded the electrical power company that became Sønnico in the capital Kristiana (the name of Oslo at that time) in 1910. Hydroelectric power had become very popular in Norway and was used for both business and domestic purposes. To begin with, the firm constructed small rural power stations and carried out the electrification of buildings before becoming involved in the production of lead-acid batteries. Initially, these were large, stationary power units, but later they were reduced in size to enable them to be used in cars and boats. The batteries were very successful, being manufactured in large quantities and sold under the brand name Sønnak for many years, and were the company's first move into retail products. The engineer K. F. Oppegaard, who would later become the sole owner and manager of the product division, became a partner in 1923, and the company opened a retail outlet in the centre of Oslo in 1929. Throughout the 1930s, the company was promoted as providing 'everything from hydropower stations to lamps'. On the industrial side the company worked on projects as diverse as installing the electrical wiring on ships and building distribution boards and transformers, whereas Sønnico's domestic division produced a variety of appliances for the modern home including electric lamps, waffle irons, refrigerators and kitchen stoves.[30]

In the spring of 1939 the engineer Rolf Hofgaard, a naturally gifted mathematician and inventor, approached Oppegaard with a proposal to develop a relay-based calculating machine for which he had already been granted a patent. A one-year contract to provide Hofgaard with financial and practical assistance was agreed, but this soon expanded beyond all expectations into a lengthy, complex and expensive endeavour that was to have serious repercussions for the company.[31]

Setting up such a project at that time was not straightforward. The outbreak of war meant that Norway was unable to gather information on technical developments from elsewhere, and so a small group of isolated but productive researchers gathered within the company and laboured away in their own small workshop trying to reach the ambitious goal of creating a fully automatic office machine. The intention was that the end result would be able to handle all the accounting-related functions a company would require, including raising orders, invoices, price lists, payroll details and so on, as well as being able to be programmed to work for different types of businesses. The necessary data for the machine was inputted manually using an electric typewriter with special contacts that drove an integrated punch tape machine. The resulting punch tape then provided a set of instructions as a binary signal to the machine to process and print the relevant accounting forms.³²

Product development continued throughout the war, and by 1946, the company had been granted six Norwegian and twelve Western industrialized country patents, including one from the United States, indicating the level of originality in the design of the machine. However, it took another ten years of laboriously involved research and development before a working prototype was produced that was able to prove that Hofgaard's theoretical principles were right. The numerous reasons for such a long delay included the understandable lack of resources within a small company such as Sønnico to support such a large-scale project and the overzealous nature of the machine constructor George Danielsen, who had slowed the project down by insisting on mastering every aspect of the computer project himself, including all the financial aspects, the engineering plans for future production and even the planned marketing campaign.³³

Between 1955 and 1958 the initial working prototype was demonstrated to a whole series of potential partners and a number of new stakeholders were brought on board. The potential market for a small business computer was a large one—it was estimated that there were around 600 such customers in Norway in the retail sector alone—but despite the prospective profits, funding to complete the project was fast running out. Technical alterations, small adjustments and mechanical improvements were constantly being made to the working prototype, using up even more of the company's limited development budget. A second, unrealized prototype was planned that was to incorporate all these modifications to a production standard, even though impartial tests carried out in 1956 concluded that the original machine was already highly cost-effective and performed better than necessary.

Eventually, the company recognized that the window of opportunity to market the Hofgaard machine had in fact disappeared. The project had run on for far too long. Electromechanical relays were already old technology when the project had started twenty years earlier. The first electronic computers developed during the war had utilized the technology of vacuum tubes that, though fragile, were considerably faster than relays. This approach had been quickly adopted by a number of competitors and had formed the beginnings of a new industry, making electromechanical computers appear outdated. The

Fully assembled prototype of the Hofgaard Machine, c. 1955.

final straw came, though, with the introduction of the silicon transistor into computers in 1954. As state-of-the-art technology that was quickly forming the basis of a whole new generation of computers, the advent of the transistor effectively meant that any further investment in developing machines based on old mechanical relays was unjustifiable. Despite having invested an incredible 1.5 million Swedish kroner by this point, the company decided to terminate the project.³⁴ In fact, the endeavour had used so much of the company's money that Sønnico had become financially unstable, and a new partner, August Cappelen, had to be brought in to invest some much-needed capital. On becoming a 50 per cent shareholder, Cappelen quickly moved to close Oppegaard's activities down, meaning that Sønnico never did get into the computer industry. In the 1960s, the discarded, redundant remnants of the Hofgaard Machine were donated to the Norwegian Museum of Science, Technology and Medicine, where they remain to this day.³⁵

Project **THE NORDSIECK COMPUTER** Designer **ARNOLD NORDSIECK**
Client **UNIVERSITY OF ILLINOIS AT URBANA-CHAMPAIGN** Date of design work **1950**

Detail from the brochure of the Analyzer Corporation, c. 1960.

DESCRIPTION **A portable desk-based electromechanical differential analyser, 76 by 152 by 96 cm (30 by 60 by 38 in.) weighing 227 kg (500 lbs). Programmable through an integrated plugboard connected to a series of internal synchro motors and mechanical integrators.**

The Nordsieck Computer was conceived as a low-cost alternative to large and complicated mechanical differential analysers, with the initial machine built from just a few hundred dollars' worth of war surplus equipment. Replicas of the machine were built and worked faultlessly in a few university departments, but the Analyzer Corporation's plans to steal the design and put it into production came to nought. By the time machines were ready, electronic digital machines were quickly reducing in price, making analogue calculators a thing of the past.

The original Nordsieck computer, 1950.

Arnold Theodore Nordsieck was a well-connected and highly respected theoretical physicist. He studied initially at Ohio State University, where he completed a master's degree before attending the University of California at Berkeley under the direction of J. Robert Oppenheimer (of Manhattan Project fame), completing his PhD in 1935. His thesis, "The Scattering of Radiation by an Electric Field," was the first step in a successful career in the emerging field of quantum electrodynamics.

With teaching positions at the time being hard to come by, Nordsieck worked for a few years on postdoctoral fellowships, first in Leipzig under Werner Heisenberg and later at Stanford, where he collaborated with Felix Bloch to produce highly regarded research papers. This was followed by a period teaching physics at Columbia University and working in their radiation laboratory and also a spell at Bell Telephone Laboratories in New York where he worked on radar development. Eventually, following the recommendation of three distinguished physicists, Nordsieck was awarded a teaching position at the University of Illinois at Urbana-Champaign (UIUC) in 1947, where he became a professor of physics.[36]

Although the latest developments in computing following the Second World War were almost all focused on the digital electronic computers based on vacuum tubes (or valves), inspired by the likes of ENIAC at the University of Pennsylvania, such machines were cutting-edge technology and consequently expensive. As a result, mechanical, analogue differential analysers of the type developed by Vannevar Bush in the 1930s were still very much in use, although these too were large and expensive pieces of equipment. Nordsieck 'became convinced that there was a need for a smaller, cheaper instrument of this type with a faster and more convenient set up procedure'.[37] Shortly after arriving at UIUC, he began work on building such a machine, the concept for which he had initially developed while at Bell Laboratories,[38] in order to model complex differential equations. Throughout the machine's development he kept a set of detailed, dated notes of his work.[39]

The mechanical differential analysers with which Nordsieck had experience used mechanical shafts and wheels. In very simple terms, the variables used in differential equations were represented in these machines by rotating the shafts and counting the number of revolutions. Purchasing $700 worth of war surplus materials, Nordsieck replaced these shafts with six mechanical integrators, six multipliers and four adders (basically these were

Arnold Nordsieck at the controls of his differential analyser, 1950.

Arnold Nordsieck with an early version of his gyroscope, c. 1954.

mechanical gearboxes which could be set at different ratios) and converted their various positions into electrical signals by using forty-three synchromotors. With these components, Nordsieck's analyser achieved similar results to machines that had 'cost other institutions as much as $100,000',[40] and he was even left with enough spare parts to build a second machine. Values for equations could be inputted by using two hand cranks on the front of the machine or through two plotting tables on the top of the machine, using the pointer to follow curves on a graph. The plotting tables could also be used as an output device, drawing the curves resulting from an equation. The output signals were made available at the master plugboard on the side of the machine, which could be wired in different configurations to solve various mathematical problems. This approach meant that the operating power required was kept to a minimum, with the whole machine being easily operated through a domestic mains electric supply.[41]

The first completed Nordsieck Computer became operational in early 1950 and was celebrated in an article in an Illinois state newspaper as 'a baby brain'.[42] The article explained that equivalent computers would fill a big room and require a great deal of power, yet Nordsieck's machine was built on a cart smaller than a desk which could be wheeled around from one laboratory to another

and used 'less electrical current than a toaster'. It also stressed the simplicity of the machine by stating, 'It can be rigged for problem solving in ten minutes while the big machines—there are five or six in this country—take several hours or a day.'

Nordsieck's analyser worked very successfully, and replicas of the machine were built and operated at Purdue University in Lafayette, Indiana, and at the radiation laboratory at the University of California at Berkeley (despite an eventful transportation of the machine involving a road crash).[43] Richard Norberg, who had worked on the original machine as a graduate student with Nordsieck, built another copy of the machine from Nordsieck's spares in 1956 after he moved to Washington University in St Louis. This particular machine was donated to the Computer History Museum in Mountain View, California, where it now forms part of its permanent exhibition.

In the late 1950s and early 1960s, the Analyzer Corporation of Los Angeles attempted to manufacture a production version of Nordsieck's analyser and produced a promotional brochure for the 'Nordsieck Computer Model C', extolling its low cost, simplicity and ease of use. Nordsieck's son, Richard, recalls, 'We were understandably excited that he might finally see some financial reward from his invention. But sadly, that was not to be.'[44] It transpired that despite Nordsieck's award of a patent for the mechanical integrating device used in the machine, the Analyzer Corporation thought it could take the design without any compensation to its inventor. Nordsieck hired a lawyer to pursue his rights under patent but soon found out that the lawyer had switched sides and chose to represent the Analyzer Corporation.[45] By this time, Nordsieck had moved on to even better things and was well into the development of a much more significant invention, the Electrically Suspended Vacuum Gyroscope,[46] for which he had successfully secured funding from the Office of Naval Research. In 1961, Nordsieck moved to Santa Barbara, taking up a position at General Motors Defense Research Laboratories, where he led a large team to improve his 'Gyro' for military and commercial navigation applications.

As it turned out, the Analyzer Corporation seems to have had little success in selling a production version of the computer. Nordsieck heard that the company had received an order from the aerospace company Lockheed but, despite its fancy brochure, nothing else. In any case, by this time electronic computers were advancing so rapidly and becoming available so widely that analogue computers were soon seen as obsolete pieces of equipment.

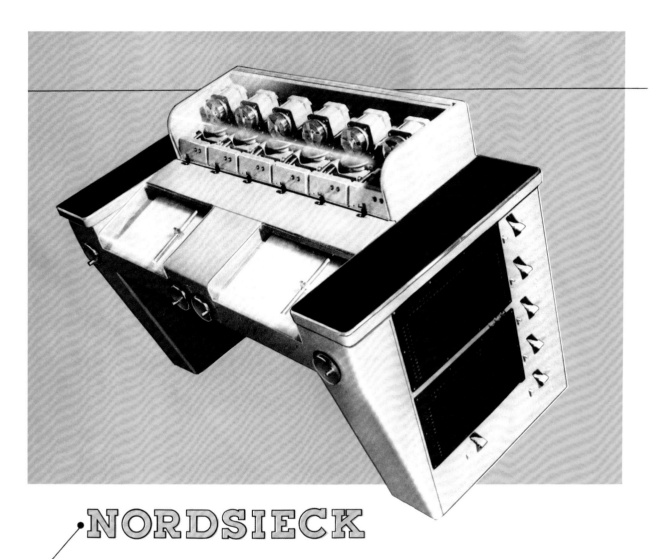

NORDSIECK

A rugged mechanical differential analyzer, geared to keep pace with your day-to-day requirements of accuracy, flexibility, simplicity.

THE ANALYZER CORPORATION

Image from the brochure of the Analyzer Corporation, c. 1960.

Project **SAAB D2**
Client **SAAB (SVENSKA AEROPLAN AB)**
Director **VIGGO WENTZEL**
Date of design work **1960**

DESCRIPTION **A desk-mounted, transistor-based minicomputer in a bespoke vertical steel cabinet. The D2 measured approximately 1.25 m by 1 m by 30 cm (4 ft by 3 ft by 1 ft) and weighed 200 kg (440 lbs). Programmed via a small, central control panel, the D2 had a memory capacity of 15 kb.**

The D2 prototype was an important milestone for the Swedish aircraft manufacturer Saab, being the first transistor-based computer to be made in Sweden. Although never manufactured as a commercial machine, it did lead directly to a whole series of commercial 'heavy line' business computers, as well as to the first transistorized computer to be completely contained within an aircraft.

Saab's background in computing stemmed from a design originally developed by the Swedish government, which was one of the fastest computers in the world at the time. This design was itself an improved copy of one of the earliest American electronic computers developed at Princeton University immediately following the Second World War.

Working prototype of the Saab D2, 1960.

Hardware designer Gösta Neovius at the control panel of BARK, 1950.

In the early days of the electronic computer industry, in 1947, the Swedish government, not wanting to be left behind in this emerging field, set aside the significant sum of two million kroner for the purchase of American machinery. A number of engineering scholars were also sent to study at American centres of computing development at the Institute for Advanced Study in Princeton, at Harvard University and at IBM headquarters in New York.

When it became clear in 1948 that the outbreak of the Cold War meant that export licences for American computers were not going to be forthcoming, the Swedish government set up a new public institution, Matematikmaskinnämndens arbetsgrupp (MNA; meaning 'working group of the mathematical machine committee'),[47] and explored the possibility of building its own computers. Conny Palm, an associate professor at the Royal Institute of Technology in Stockholm, had already planned to build a drum-based electronic computer for the military, as well as a relay-based electromechanical machine that could be developed more quickly if required. MNA decided to build this relay-based machine, which became known as BARK (binär automatisk reläkalkylator, or binary automatic relay calculator).[48] The construction team, led by Palm, used 8,000 telephone exchange relay switches in its design, which was programmed through the use of a plugboard.

BARK was only ever intended to be an interim measure, and as soon as it was completed in April 1950, work immediately began on a much larger electronic computer called BESK (binär elektronisk sekvenskalkylator, or binary electronic sequence calculator). This project was led by Erik Stemme, one of the scholars who had been sent to study at the Institute for Advanced Study (IAS) in Princeton,[49] and the final design was basically a direct copy of the IAS machine built there between 1945 and 1951 under the direction of John von Neumann. It covered an area of 30–40 sq m and contained 2,400 vacuum tubes or 'valves' and 400 diodes,[50] and as a result of improvements on the design of the IAS machine made by Stemme, when it became operational in 1953 it proved to be one of the fastest computers in the world.[51]

BESK was used for a wide variety of data-handling purposes, and as was normal practice then, time on the computer was rented out to those institutions that required its capabilities. Calculations were carried out on BESK for the Swedish meteorological agency, telecommunications companies, the Swedish nuclear energy industry and the Swedish nuclear weapons program. It was also used by the road authority for the design of road profiles and even by the Swedish National Defence Radio Establishment to decode intercepted encrypted Russian messages in the Cold War; but it was its use by Saab, the Swedish aircraft manufacturer, which proved most fruitful for the machine's future.

The cost of mainframe computer time was prohibitively expensive to many companies, and obviously non-exclusive, meaning it was not always available when required. Saab's requirements for computing time were such that it made it worthwhile for the company to invest in developing its own version, with an even greater capacity than BESK, to be used in the design of the next generation of Saab military aircraft.

Rather than start development from scratch, Saab was granted a licence from MNA in 1954 to produce one copy of BESK. The new machine was to be twice as fast as BESK and was called SARA (Saab's räkneautomat or Saab's computing automaton). It was completed in 1957 at Saab's headquarters in Linköping and installed in the Calculation Department, along with its dedicated magnetic tape storage system called 'Saraband'.

Sweden's first electronic computer, BESK, c. 1956.

In the end, the technology within BESK was widely disseminated, and a number of other companies made similar copies of the machine. One such machine called SMIL (Siffermaskinen i Lund, or the digital machine in Lund) was installed in 1956 at Lund University and was their main computer for over fifteen years,[52] while in Denmark, Regnecentralen built the first Danish computer, DASK (Dansk algoritmisk sekvens kalkulator or Danish algorithmic sequence calculator). This began as a straight copy of BESK, but various improvements made it a different and much larger machine weighing 3.5 metric tons. Another Swedish firm, Åtvidabergs Industrier AB (makers of 'Facit' mechanical desk calculators), recruited a number of people from MNA and produced a BESK copy called Facit EDB (electronic data behandlar, or electronic data processing) in 1957. The visual similarity between the console of the original BESK and the Facit EDB is striking, with many identical components. This was presumably done deliberately to provide a level of visual continuity, even though the smaller size of the transistors used in the Facit EDB meant that, internally, the console consisted mainly of fresh air.[53]

In the same year SARA was competed, the Missile Division of Saab began work on a missile project called 'Robot 330'. The two-stage supersonic missile was intended to carry a nuclear payload to strike against the Baltic harbours in the case of a Soviet invasion of Sweden. Using the experience gained in the construction of SARA, a small transistorized computer was planned to provide the navigational control of the Robot 330 missile. This project, under the direction of Viggo Wentzel, was code-named SANK (Saab automatiska navigations-kalkylator, or Saab automatic navigation calculator). When the missile project ceased, Saab realized the

SARA, Saab's first binary computer system, 1957.

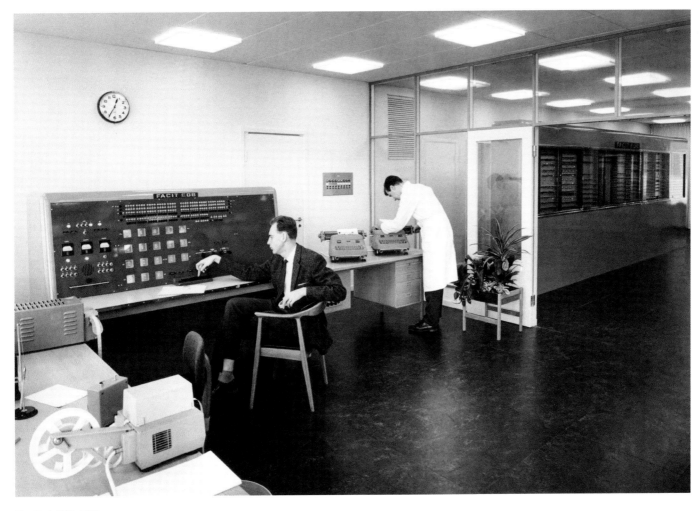

The Facit EDB, 1957.

potential to sell the same technology in a commercial context, and a project named D2 was started, aiming to produce a business version of the computer.

By 1960, a working prototype of the D2 had been produced (see main image) which weighed 200 kg (440 lbs) and could just about fit onto a desktop. It was the first transistorized computer built in Sweden and had a memory capacity equivalent to about 15 kb. It was a significant achievement, raising interest (and orders)[54] from potential clients, but never went into production. Instead, along with elements from SARA, it directly informed the development of a number of larger, 'heavy line' business computers. In terms

The Saab D21 installed at the electric power company Skandinaviska Elverk in 1962.

The CK37 (Centralkalkylator) for the Saab 37 Viggen aircraft, 1971.

of internal hardware, the D21 of 1962 was very close to the D2 but with the addition of more memory and a slightly different architecture, a paper tape reader and a magnetic tape station. A total of thirty-two of this particular computer were sold. This was followed by the D22 in 1968, the architecture of which was influenced by the IBM System/360. This was a much more successful machine, seventy-one being sold under the brand name 'Datasaab'.[55] The final version, the D23, was based on semiconductors. It was announced in 1972, but manufacturing problems meant that only four were produced before the commercial computer business became part of a new company, Saab-Univac. This in turn became Datasaab AB before being acquired by Ericsson Information Systems in 1981.

Another interesting line of development from the SANK/D2 project was an investigation into the possibility of miniaturizing a computer to fit inside a military aircraft to aid with navigational tasks. However, the level of miniaturization required to place a complete computer inside an aircraft took longer to achieve. Numerous prototypes were developed and tested on the ground and in flight between 1963 (the NSK computer) and 1968. Finally, the CK37, the world's first airborne integrated circuit computer, was completed in 1971 and successfully used in the Saab 37 Viggen aircraft until the 1990s.

Project **HONEYWELL KITCHEN COMPUTER** Designer **DON KELEMEN**
Manufacturer **HONEYWELL COMPUTER CONTROL DIVISION, US** Date of design work **1969**

DESCRIPTION **A free-standing minicomputer in a futuristically styled fibreglass pedestal unit. Programmed through a series of switches, with a single-line numerical display behind a smoked plastic 'windscreen' strip. Dimensions: 84 by 64 by 109 cm (33 by 25 by 43 in.).**

The Honeywell Kitchen Computer is described in a number of places as a piece of fantasy—an imaginary computer that was advertised for sale but never sold. To a certain extent this is true—the idea that anyone would actually buy a computer that cost $10,600 merely to store recipes (at a time when that amount could have bought a number of new cars or a house in suburbia) is indeed fantastic. Mind you, the incredible price tag included a two-week computer programming course to train the lady of the house how to enter recipes into the computer using only the input switches and read them from the octal (base 8) display (there was no text-based keyboard or monitor).

It is now clear that the Honeywell Kitchen Computer was actually a publicity stunt, continuing a long line of fantasy gifts offered by the upmarket American department store Neiman Marcus. It appeared in their 1969 Christmas catalogue and in magazine advertisements along with text that stated, 'If she can only cook as well as Honeywell can compute'.

Image of the Kitchen Computer from LIFE *magazine, 12 December 1969.*

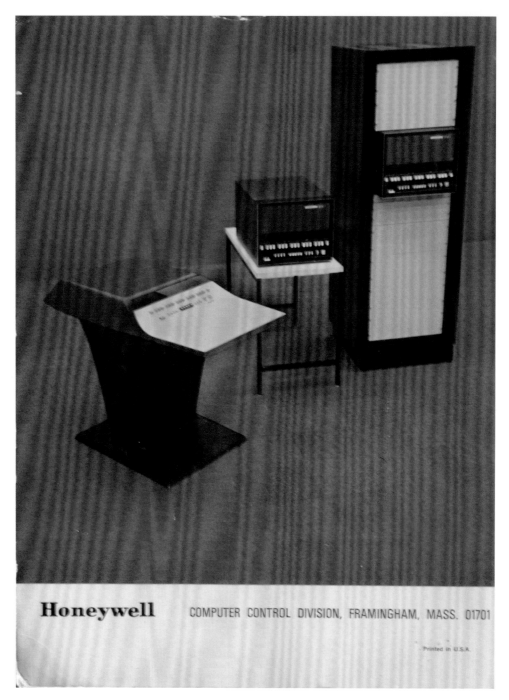

Back cover of Honeywell's manual for the H316 minicomputer, 1969.

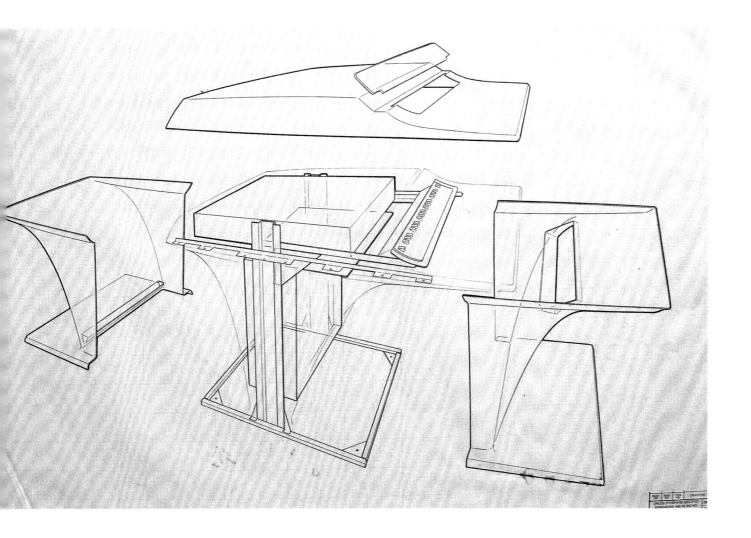

Don Kelemen's exploded drawing of the H316 minicomputer pedestal unit, 1969.

Honeywell advertisement for the H316 minicomputer, Datamation, *May 1969.*

Neiman Marcus catalogue for the Kitchen Computer, 1969.

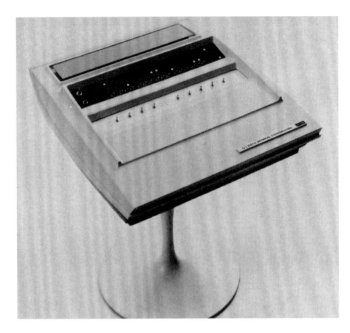

Data General Nova minicomputer, featured in Datamation *magazine, November 1968.*

Although the Kitchen Computer was a 'spoof' product, it was based on a real computer by a real computer company—the Honeywell H316 pedestal computer. This was one version of a minicomputer from Honeywell's 'Series 16' range of the 1960s. In 1966, Honeywell bought Computer Controls Corporation (CCC), renamed it Honeywell Computer Control Division and in the process took on two of its employees, the computer engineer Gardner Hendrie and the industrial designer Don Kelemen. It was Hendrie who developed the Series 16 technology from his earlier design of the world's first sixteen-bit minicomputer, CCC's DDP-116, whereas Kelemen created the radical design of the pedestal version.

Kelemen was responding to a marketing brief to add a "wow" factor to Honeywell's H316 minicomputer in order to allow it to compete with a futuristic offering from its competitor, the pedestal-mounted version of Data General's Nova minicomputer. The unique form of the H316 pedestal was achieved by encasing the components of the standard version into a fibreglass shell, resulting in a futuristically styled, red, white and black unit that looked as if it could have been taken straight from the set of *Star Trek* or *2001: A Space Odyssey*. As with the standard rack-mounted or table-top versions, commands were entered line by line in numerical code through a set of switches, here placed towards the back of the white desk surface. The only display was a series of illuminated figures behind the smoked plastic 'windscreen' strip that showed the entered commands not in text, but in octal (base 8) numerals.

Promoting itself as 'The Other Computer Company', Honeywell, in its advertisements for the H316 range, used fashionably blurred imagery and focused on the science fiction aspect of the pedestal version under the futuristic heading 'Ultramation'. The product sold very well in its standard configurations, but the advertisements hint that the pedestal version was not to be taken too seriously, as they stated that customers would 'probably want rack-mountable or table-top models more than the glamorous version pictured here'.

By the time the Honeywell Kitchen Computer first appeared in the 1969 Neiman Marcus Christmas catalogue, the chain had a long-standing tradition of offering extraordinary and extravagant items in its Christmas catalogue, initiated by the founder's son, Stanley Marcus. In an attempt to gain free media coverage, it occurred to Marcus to create hugely expensive gifts and put them in the catalogue. The first of these, in 1960, was a matching pair of 'His and Hers' Beechcraft airplanes for $176,000, and the story was reported worldwide. Later gifts included a real Black Angus bull with a sterling silver barbecue cart, authentic Chinese junks and a made-to-order Noah's ark. Genuine Egyptian mummy cases offered one year made front-page news when one was unexpectedly found to contain a real mummy. Marcus wrote, 'The Christmas "idea" pages evolved for the most part into my area of responsibility, for I seemed to have a better understanding than most of my associates about which ideas could be depended upon to produce favourable news coverage and which could not. I have always tried to select concepts which have a degree of credibility (however faint), are not hackneyed, and will evoke the question "I wonder if they really sell any?"'[56] The Kitchen Computer promotion worked particularly well, being picked up by various news channels and magazines, including *Esquire*, *Playboy* and *LIFE*.

How much impact could a computer that was advertised as a fantasy gift have on the serious development of computing technology? Surprisingly, the answer is quite a lot. Gordon Bell, vice president of engineering at Digital Equipment Corporation (DEC), wrote a memo to the company's Operations Committee, citing the Neiman Marcus advertisements as the catalyst for his thoughts about the potential of computers in a domestic setting.[57] Bell's view was that even though the Kitchen Computer as proposed by Neiman Marcus was 'useless', the use of computing technology in the home was inevitable, and so DEC should be at the forefront of developments. The memo described computers in the home being used for a range of educational, organizational and leisure purposes; predicted that in the future, computers could control heating, air conditioning, lighting, burglar alarms, dishwashers and music systems; and saw that one day, computers would play complex games such as chess, send and print letters over a network and be used for shopping, printing newspapers, ordering books and even 'scanning periodicals for keeping informed as to what to read'—all now everyday functions of home computers.

So the H316 pedestal/Kitchen Computer did have some influence. As a real product, its science-fiction-inspired form enabled Honeywell to promote itself as a progressive company, to differentiate itself from its mainstream competitors and to align itself with younger, innovative companies such as Data General Corporation. As a fantasy product, the Kitchen Computer created a huge amount of publicity for Neiman Marcus. In addition, it inspired those working at the forefront of computer developments to realize that, despite the limitations of technology at the time, there was real value in seriously considering a domestic market for computers.

There is a physical example of the Kitchen Computer in the Computer History Museum in California, which for many years was thought to be purely a prototype for demonstration and promotion purposes. As it turns out, it is one of a very small batch of the pedestal version of the H316 produced to meet requests from business customers that were taken with the futuristic design.[58] Don Kelemen confirms, 'We built the computer as a marketing tool, but to our dismay we received several orders. I believe that we ran 20 of the shells and frames. The rest were standard parts off the desktop version. As far as I know none were sold as "Kitchen Computers" but the publicity didn't hurt.'[59]

Neiman Marcus advertisement for the Kitchen Computer, 1969.

Project **CTL MODULAR THREE MINICOMPUTER** Designers **BILL MOGGRIDGE AND JOHN ELLIOTT, MOGGRIDGE ASSOCIATES**
Client **COMPUTER TECHNOLOGY LIMITED, UK** Date of design work **1972–1973**

DESCRIPTION **The Modular Three (M3) Minicomputer concept was approximately 46 cm (18 in.) wide, 56 cm (22 in.) deep (front to back) and 15 cm (6 in.) high. Built on a chassis for rack-mounted use, the main body of the 'desktop' version was made from folded aluminium sheet, with structural foam mouldings used for the side ventilation panels. Input controls were housed on an acrylic touch-sensitive control panel.**

Moggridge Associates' design work for the CTL M3 minicomputer was intended to replace the company's existing product line, the 'Modular One' rack-mounted minicomputer system, which was proving expensive to manufacture.

Much smaller than its predecessor, the front control panel of the proposed design was an innovative, removable touch-sensitive acrylic panel, which was connected to the processor unit via a coiled cable. This concept was intended to allow versatility in use as, when used as a desktop unit, the control panel could be used either in situ (as shown) or be pulled forward to rest on the desk surface, or it could be turned vertically to fit across the front of the unit if it was rack-mounted. In this way, it could also be used as an extension to the 'Modular One' product line. Unfortunately, internal company politics meant the design never went into production.

Prototype of the M3 minicomputer desktop unit by Bill Moggridge, 1973.

The Modular One installation at the Applied Psychology Unit (APU) in Cambridge, 1970.

Computer Technology Limited (CTL) was founded in 1965 by Iann Barron, an ex-employee of Elliott-Automation. Barron had some radical ideas for an innovative minicomputer system but could not persuade Elliott-Automation to incorporate them into its next generation of computers. Frustrated, he decided to raise venture capital, go into business and manufacture computer systems himself. Originally consisting of just five design and development engineers, the company grew rapidly, doubling its staff, accommodation and turnover each year for the next five years and becoming 'one of the fastest growing computer manufacturers in Britain'.[60]

CTL's first product was the 'Modular One', a sixteen-bit minicomputer announced in 1968. The Modular One was unusual in that it was, as the name suggests, a modular system of individual units, each fulfilling a different function. These units, each about the size of a washing machine and fitting into an industry-standard 19-in. rack system, could be purchased and assembled in a variety of ways depending on the customer's requirements. This resulted in almost every Modular One system installation being completely different, which had serious implications for efficient maintenance. It also proved to be an expensive way to produce computers. Even after a cost-reduction exercise in the mid 1970s, the Modular One could not compete with similar but cheaper products from DEC and Data General.

The Modular One never sold well outside of the United Kingdom, yet despite its expense, British scientific institutions were significant customers for CTL. The system's size, cost and unreliability, though, soon became apparent. An early customer for the Modular One was the Medical Research Council Applied Psychology Unit (APU) in Cambridge. Their system was installed in 1970, and one of its early uses was to prepare speech materials for use in experiments.[61] The modular design meant that a moderately large room was required to accommodate all the hardware, even though it had just 24 kb of random access memory (RAM).[62]

A Modular One installation purchased in 1973 by the University of Birmingham had just 16 kb of main memory and despite numerous discounts cost £23,000. It was mainly used for research into programming languages and operating systems but suffered from poor reliability, breaking down three or four times a year. On one occasion, a disk fault meant the machine was inoperable for over two months while the engineers struggled to diagnose the fault.[63]

From the time of the Modular One's launch, CTL's Board of Directors, which included venture capitalists and others with little knowledge of computers, pressed Iann Barron to develop the next generation of computer architecture. Barron stated, 'I was completely opposed to this, because I knew that we could not afford to obsolete our investment in software by creating a new architecture. I did manage to contain this pressure and develop a hardware enhancement—the Modular Two (visually identical apart from being in orange coloured units)—but eventually I lost control and was required to design a completely new architecture, the Modular Three.'[64]

Barron retained the services of Moggridge Associates to develop design concepts for the Modular Three (M3). Moggridge recalled that 'CTL was a very exciting client to have in those early years, as Ian Barron was high profile and ambitious'.[65] Being at the time a recently founded consultancy of just three associates, their relationship with clients was paramount. Through his experience of working on a regular basis for Hoover, Moggridge had developed an arrangement he called 'partially in-house

The Modular One installation at the University of Birmingham, 1976.

Detail of the M3 'floating keyboard', 1973.

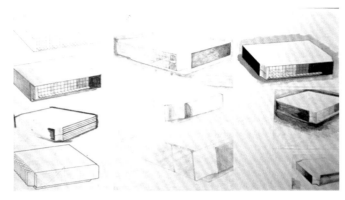

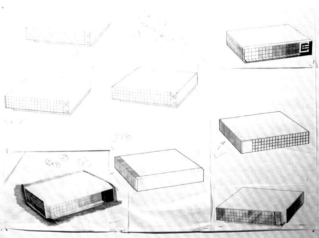

Early concept sketches for the M3 project, 1972.

consultancy' which involved working within the client's premises for a particular period each week throughout the design and development stages and continuing until the product reached production. This gave both sides benefits: for the consultants it allowed the freedom to be self-employed and work for other clients, while for the clients, it meant there would be hands-on advice when it was required to solve any teething troubles with designs. It also meant that the designers could keep stricter control and ensure their design work was not unduly interfered with by in-house engineers. This successful client relationship meant that Moggridge, along with his colleague John Elliott, got to know everybody within the company and the ins and outs of the technological issues involved in their products.[66]

The concept of the M3 was a high-performance replacement for the Modular One and Modular Two. The pace of technological development since the Modular One system had been developed just a couple of years earlier was such that a much smaller, self-contained unit was now feasible. Each washing machine–sized module of the Modular One system was replaced by a single printed circuit board, which meant a whole computer system could be reduced to a single box.[67] In fact, 'the M3 minicomputer would have been one tenth the size of contemporary competitors'.[68] In order to permit the new generation of products to work alongside the original Modular One products, the units were designed to fit a standard chassis that would allow it to be rack mounted. To give greater flexibility and to allow the units to be used in different settings, versions were also designed to be stand-alone products that could be used on a desktop.

Moggridge's final design had an epoxy-powder-coated sheet steel casing, with structural foam mouldings used for the side panels to allow the internal components to be ventilated. The unit had a unique removable control panel or 'floating keyboard' that could be used in different configurations depending on the computer's location. This panel, made of finely sand-blasted bronze acrylic within an aluminium pressed frame, allowed touch-button programming and contained LED indicators and a backlight to light up the graphics.

Between them, Bill Moggridge and John Elliott developed a whole series of different concepts for Barron during the course of their consultancy, including a full range of rack-mounted peripherals and integrated furniture. Unfortunately, the grand scheme for a new generation of CTL products was never realized, despite the M3 processor having speeds of performance that were 'extremely

An alternative desktop design for the M3 minicomputer by John Elliott, 1973

fast for its era'. The speed was due to a radical, patented circuit design, which, although potentially highly lucrative, Barron chose not to enforce. Barron's relationship with the Board of Directors continued to deteriorate until 'finally, I was dismissed, and the project was (rightly) abandoned, and all future enhancements were to the original architecture'.[69]

After Barron left Computer Technology Limited in March 1975, he worked as a consultant to Intel, Motorola and Texas Instruments before founding the semiconductor company Inmos with £50 million of financial investment from the UK government's National Enterprise Board. During this period, he developed his ideas for a revolutionary parallel-processing microprocessor called the 'transputer'. During the late 1980s the transputer was considered by some to be the way forward for computing. Although this ultimately proved not to be the case, the transputer design was highly influential and inspired a number of designs from others, the descendants of which are still used today. The government sold its controlling share of Inmos to Thorn-EMI in 1984. CTL continued to work in the computer industry by outsourcing computer systems and rebadging them before being taken over by Applied Computer Techniques (the manufacturers of Apricot computers) in 1989.

Personal and Portable Computers

It is hard to believe, in this age of ubiquitous, interconnected, personal mobile computing—where millions of people regularly carry with them large amounts of computing power, possibly without even realizing it—that at one point in time, the very idea of an individual owning his or her own computer was seen to be completely ridiculous. It is true that there were some isolated visionaries who predicted the day that people would have their own machines,[1] but these were very much in the minority. By and large, until well into the 1980s, the established players in the computer industry could see absolutely no point in developing such a device. Who would use such a thing, and what would they use it for?

This restricted view of what a computer could be lay behind the often recounted and out of context remarks by people high up in the computer industry, which given the luxury of hindsight sound so ridiculously uninformed today. Thomas Watson, the chairman of IBM, is said to have stated in 1943 that there was a world market for maybe five computers, but this was when huge electromechanical computers filled whole rooms and cost millions of dollars and more than twenty years before the computer scientist Gordon Moore realized that computers were constantly getting smaller and increasingly more powerful and predicted that they would continue to do so at a constant rate.[2] Similarly, Ken Olsen, the co-founder of DEC Computers, told the World Futures Conference in 1977 that there was no reason for any individual to have a computer in the home (although it is argued that he was not talking about 'home computers' but about extremely powerful computers that would control everything in the home from the lighting and heating to the entertainment and meals). Whatever the truth or otherwise of these statements, they reflect an insular industry attitude towards individuals owning computers—one that was directly responsible for a number of attempts to create personal computers being carried out almost as a clandestine activity.

Whereas the trajectory of the development of the office computer had its beginnings firmly in the corporate ethos of the business machine, the appearance of the personal computer as a commercial product was largely a result of groups of enlightened individuals interested in computer technology for its own sake. The DIY electronics fanatics who had spent all their spare time building home radios found a whole new world opening up to them when the microprocessor became available, and many built their first computer by assembling mail-order kits of parts, despite there being very little that they could actually do with them once they were built.[3]

The social aspects of this hobbyist approach were important, with like-minded people forming clubs and arranging meetings where they could get together and help each other to build computers or suggest improvements to each other's circuit designs.[4] At one such club the seeds were sown for what would one day become the world's biggest company.[5] Steve Wozniac, an experienced radio ham who worked by day designing calculators for Hewlett-Packard, and his friend Steve Jobs (who at the time worked at the computer game company Atari) had for some time worked together on electronics projects, including the now infamous 'blue box' which allowed illegal phone calls to be made free of charge. They started attending meetings at the Homebrew Computer Club in Palo Alto, talking with others intent on producing their own microcomputers. In 1975, many of the club's members had built one particular computer, the Altair, from a kit sold by MITS Ltd. Wozniac took the circuit apart and was convinced he could do it better and cheaper, and the Apple computer was born. Seeing the potential for the computer to sell in big quantities, the pair offered the rights to both their employers, Hewlett-Packard and Atari. Both of them turned it down.[6]

From this inauspicious start, the form of the home computer quickly moved from being a handmade wooden box of soldered parts to being a sought-after shop-bought appliance. By 1977, there were three clear front runners dominating the home computer market—the Apple II, the Commodore PET 2001, and the Tandy TRS-80—and if the advertisements are to be believed, home computing moved from the hobbyist's confines of the garden shed or spare bedroom to take centre stage in the home.

As the cost of computing power became ever cheaper and the market for home computers ever larger, there was a proliferation of competition. Without the requirement for huge corporate research and development resources, readily available, off-the-shelf components meant that anyone could now set up a computer business on a shoestring, and many did just that. In the late 1970s and early 1980s, thousands of small computer companies emerged, many little more than one-man bands. Numerous new products were launched each week in a desperate attempt to take control of even a small part of what was rapidly becoming a huge market. Many of these were fairly mundane rehashes of existing technology but nevertheless presented the consumer with a bewildering array of variety in form and colour. Unfortunately, most of them disappeared just as quickly as they had arrived. Over the course of just a few short years, the huge proliferation of different designs started to disappear as one particular personal computer started to dominate. In the office, computers had changed too. Gone were many of the centralized computer rooms, operated through remote terminals on a timeshare basis. Instead, smaller distributed computer systems started to appear, first built directly into office desks and later as stand-alone desktop machines. The most popular of these, the IBM PC of 1981, was to some extent a hasty response to the growing popularity of the Apple II—a home computer that had become desirable as a business tool because of the spreadsheet software VisiCalc and that was a machine threatening to infringe upon IBM's office equipment territory. With the might of a huge base of installed machines and a well-established supporting infrastructure of third-party software behind it, the IBM PC was unstoppable. Unfortunately, in order to reach the market as quickly as possible, IBM had utilized many readily available parts and a very open architecture easily copied by others. Soon afterwards, almost every computer launched took the safe option of being an IBM compatible clone—almost indistinguishable from each other in form and function.

This particular decade then—from the mid 1970s to the mid 1980s—saw some huge shifts in the computer industry which were directly or indirectly responsible for many examples of vapourware. First was the move from the accepted view of a computer as a large, expensive and centralized corporately controlled machine accessed by a large number of different users to the model of the computer as a relatively cheap, distributed, individually used piece of office equipment, and second was a corresponding change from the view of the home computer as a DIY-led piece of hobby equipment requiring specialist knowledge to an increasingly common, if not yet essential, appliance for the home that could be used by all. Both of these shifts led to a massive increase in the number of computers manufactured and a wide variety of different kinds of computer appearing in the home and the office. Consequently, there were many attempts by numerous companies to enter new markets with new products, with varying degrees of success.

Project **IBM SCAMP DESIGN MODEL** Designer **TOM HARDY**
Client **IBM** Date of design work **1973**

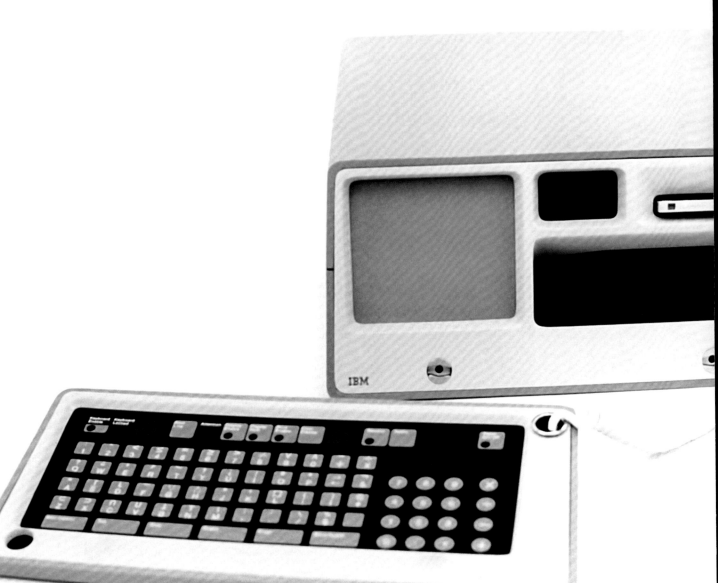

Mock-up of the IBM SCAMP 'Design Model', 1973.

DESCRIPTION **Portable computer in plastic moulded case with removable flat-panel keyboard, 6-in. CRT and data cassette drive.**

The Special Computer APL Machine Portable, or 'SCAMP', was one of IBM's earliest attempts to put full computing power into the hands of single users rather than multiple users accessing a mainframe. SCAMP was a six-month project undertaken by IBM's General Systems Division (GSD) in 1973 and resulted in a working engineering prototype which could be used as a desktop calculator or an interactive programming device, or it could run various pre-programmed applications stored on cassette.[7] The prototype is now a part of the Smithsonian Institution in Washington DC, where it is described as 'the world's first single-user computer'.

Tom Hardy produced a 'Design Model' of SCAMP to allow executives to take it with them to show potential customers how a single-user computer would look, but although based around existing technology within IBM, it never saw the light of day in this form. However, the SCAMP project itself did lead to two very successful IBM machines that bore some resemblance to Hardy's design proposal.

The working 1973 prototype of the IBM SCAMP now at the Smithsonian Institution.

Project SCAMP began in December 1972, when Dr Paul J. Friedl, an engineer of Czech descent working at the IBM Palo Alto Scientific Center, received a call from the IBM GSD in Atlanta.[8] An IBM director, Bill Lowe (who went on to become vice president), had requested that the research centre, which was working on machines using high-level programming languages, develop a scaled-down version aimed at single users.[9] This now seemingly straightforward request was a highly unusual proposition at the time, as it was then the usual practice for multiple users to access a single, large mainframe through numerous 'dumb' terminals. However, Lowe was keen to see if it was at all possible to achieve a useful level of computing capability at a small scale. Friedl was asked if he could come up with a microcomputer, 'maybe something similar to a hand-held calculator',[10] using

APL[11]—a programming language using 'symbols and common mathematical notation'.[12] Friedl was keen to work on the project and said he would put a proposal together over the next month.

Such a machine would require a portable display, a keyboard, a printer and a small magnetic storage device. The technology needed didn't seem to be ready. Working with his chief programmer, Pat Smith, Friedl based his proposal around some existing technologies from within IBM and other equipment from outside IBM. He based the architecture around the IBM PALM (Put All Logic in Microcode) microcontroller, which was designed to control different elements of a microcomputer and would allow the use of an integrated TV display. He proposed using a standard IBM keyboard, 64 kb of random access memory (RAM) (powerful for the time) in the form of four 16 kb memory cards and a new input/output card that would allow the computer to connect to a printer, a cassette recorder and the keyboard.

In January 1973, Friedl took sketches of his portable computer proposal to show to the executives in Atlanta. When he was asked how long it would take to build a working prototype, for some reason he couldn't later figure out, he said 'six months'.[13] This was going to be a lot of hard work. Friedl enlisted the help of a talented IBM engineer Joe George and a small team of engineers and programmers at IBM's Advanced Systems Development Laboratory in Los Gatos, and the SCAMP prototype started to take shape. The project was an intriguing technical challenge to Friedl, but from a marketing perspective, he started thinking, 'Why anyone would want such a thing? What possible use could a manager have for a portable microcomputer?'[14] It became clear to him that if executives who had never used a computer before were faced with using this machine, then it ought never to leave them facing a blank screen but always present them with a list of options to take. It needed menus that could be operated by pushing a single button. The capabilities of the machine also needed to be pre-programmed so that it could be easily operated, and so Friedl developed applications including project planning and financial analysis using possibly the first ever electronic spreadsheet.

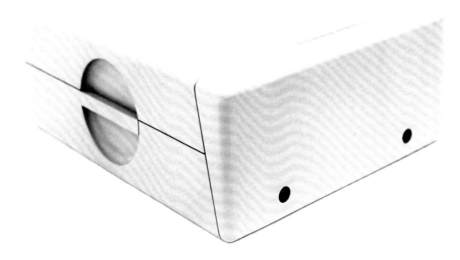

The IBM SCAMP Design Model assembled for transit, 1973.

66   The IBM 5100 Portable Computer, 1975.

The finished prototype had the internal memory cards, a full-sized keyboard, an audio cassette drive and a small 5-in. CRT display. The monitor was housed in a hinged section that could be pushed down into the body of the machine, and a cover could be slid over the keyboard to create a flat surface when closed. Jerry Garvis, an industrial designer at the Los Gatos laboratory, had created a special wooden case with a built-in carrying handle to protect the unit in transit,[15] and almost six months later to the day, Friedl took his working prototype to Atlanta. The executives were highly impressed with the product, and Friedl ended up demonstrating the machine on over a hundred successful occasions, eventually presenting it to the president of IBM, John Opel, who allowed development of the technology to go ahead.[16]

The designer of the SCAMP Design Model, Tom Hardy, was a recent graduate of industrial design at Auburn University who joined IBM in 1970. Initially based in New York, he was transferred in 1973 to Boca Raton, Florida, to join the new Entry Level Systems (ELS) group—what was to become the PC division of the company. Hardy's manager, Joe Talerico, had been to Atlanta to meet Bill Lowe and had seen the prototype SCAMP computer that Friedl had developed. He knew that the prototype was just that— never itself intended to go into production but made to prove the point that a scaled-down machine using APL could be affordably achieved. Talerico believed that IBM's executives would be highly interested in a production version of SCAMP and asked Hardy to produce a design based on the engineering prototype.[17]

The Design Model of SCAMP was intended for IBM executives to carry around to start promoting the concept of a single-user computer to important clients. The size of the case and the weight of the unit were largely determined by the selected components from the engineering prototype, but instead of the standard IBM keyboard, it used a flat-panel, touch-sensitive keyboard developed by an IBM engineer and a larger, 6-in. CRT display. The large black slot on the front was an acoustic coupling device for a telephone handset to allow the computer to connect to the telephone network. Above the acoustic coupler were a high-density data cassette drive and a speaker. The flat-panel keyboard was connected to the main unit by a retractable cable, and when in transit, the keyboard itself was to be connected to the front of the main unit by magnets.[18]

The SCAMP Design Model was shown to executives and they took the model and photographs around with them, but in this form, it never went any farther than the initial mock-up.

The technology within SCAMP, however, did lead directly to a production design—the IBM 5100 Portable Computer, which bears similarities to both the engineering prototype and Hardy's design. This machine, first announced in September 1975, weighed approximately 23 kg (50 lbs) and was specifically aimed at 'engineers, analysts, statisticians and other problem-solvers'.[19] It was sold as a 'portable' machine, but its weight meant it was not really that portable at all, and it was not designed with a handle or carrying case. It had a built-in CRT display but a 5-in. version, smaller than the one on Hardy's design. The available memory varied from 16 kb to 64 kb depending on the model, and it sold for between $8,975 and $19,975—way out of the reach of private individuals. It could be switched to be programmed in either BASIC or APL and recorded and stored data on special high-density magnetic tape cartridges. Cartridges with pre-programmed applications were available in three 'Problem-Solver Libraries' for mathematics, statistics and financial analysis.[20]

A very similar looking follow-on machine aimed at general business users rather than scientists or engineers was launched in 1978. The IBM 5110 came with programs to help with accounting, sales analysis, inventory cost reduction, growth planning and resource scheduling. Both machines were successful and sold well until, following the launch of the IBM PC in August 1981, they were both withdrawn from sale in March 1982.[21]

The IBM 5150 PC was a logical conclusion to this line of computer development, as it 'took the features and characteristics of all these early small computers and provided them in one incredibly utilitarian machine'.[22] In terms of lineage, SCAMP turned out to be 'the grandfather of the IBM PC',[23] one of the most successful computers of all time. Fittingly, given his work on the SCAMP Design Model, Tom Hardy also produced the award-winning industrial design of the IBM 5150 PC.

The IBM 5110 Computer System, 1978.

The IBM 5150 PC, 1981.

Date of design work **1976**

Client **IBM**

DESCRIPTION **A self-contained computer in an injection-moulded case with full a keyboard, a slot for data storage cartridge and a small built-in printer. Display was achieved by connecting the computer to a domestic television. Dimensions: 445 by 205 by 100 mm (17.5 by 8 by 4 in.).**

The Yellow Bird Personal Computer was a design proposal made purely to present to IBM's executive directors in an effort to convince them that the future of computing lay in the personal computer, rather than in large business computer systems.

For the leading computer company in the world, which had gained its dominant position by developing and building exactly such large-scale systems, the idea of a small computer for individual use was a difficult concept to accept. Even though the team that developed the proposal was able to prove its technical feasibility, convincing the board that such a machine should form a part of its core business was a different matter altogether.

Studio photograph of the IBM 'Yellow Bird' prototype, 1976.

In the mid 1970s, the computer industry was dominated by a small number of large corporations, of which IBM was by far the largest. Firmly rooted in the development of large-scale business computer systems that would be operated by numerous people through multiple terminals, the idea of a small, stand-alone, self-contained computer intended for use by a single individual did not sit comfortably. Some of IBM's small systems were operated by single users, but these were still serious business computers. It is true that Bill Lowe, the head of the General Systems Division, had previously pushed the development of a prototype small single-user computer—the IBM SCAMP in 1973—but this was undertaken to see if such a machine was at all technically feasible. It was, in any case, seen as only of possible interest to a very small, select group of executive users given that it would, by its nature, be prohibitively expensive. It was still a far cry from a mass-produced, low-cost personal computer.

The SCAMP prototype had, in fact, proved to be highly influential, leading directly to groundbreaking products such as the IBM 5100 Portable Computer of 1975—although this too was a specialist, expensive piece of equipment originally intended for more scientific use. Lowe had clearly proved that the technology from large computer systems could be scaled down to suit smaller machines for single users, yet despite this, the executive board remained unconvinced of the mass-market potential for such a machine. As a consequence, personal computers did not form a part of the IBM business plan.

Still, Lowe remained of the opinion that the future of the computing industry lay in this direction, and he wanted to be able to show those in power above him what a future might look like where everyone, not just business users, would have a personal computer. To further his cause, Lowe asked IBM designer Tom Hardy to start developing designs around the personal computer concept that he could 'take around and use for evangelizing the idea'.[24]

The first of these concepts was named 'Yellow Bird'. It was a small bright-yellow computer intended to be connected to a domestic television set to use as a display—a low-cost option chosen because of the high price of CRT monitors and consequently taken up by numerous manufacturers of home computers. The processor unit was housed in the rear part of the casing, and data was stored on the same type of high-density magnetic tape cartridge used in the IBM 5100 Portable Computer, which was inserted into a slot in the top of the casing. It also used the same type of red 'paddle' on/off switch as the 5100 and had the same full IBM keyboard with the addition of a separate numerical keypad and function keys. Interestingly, it had a small, built-in printer with a rotary platen that used rolls of paper like a cashier's till.

Lowe had photographs of the Yellow Bird taken in studio settings, with Hardy's daughter acting as a young potential user, shopping for the computer with her mother in a retail store and using the computer with a television in a sitting room—an unusual sight at the time. The fact that it was brightly coloured and portrayed in a domestic setting was a deliberate attempt to make it clear it was intended to be a product bought for use in the home rather than the office. Lowe used the images for 'show and tell', to try to convince various executives that in the future, the personal computer would be a serious market, but it was an uphill struggle. The executive board decided not to back the proposal.

The executive board's position on the personal computer may seem to have been overly cautious, but it has to be regarded in the context of the computer industry at that particular moment. Although individual computers had been imagined a number of years earlier,[25] actual home computers had only just emerged and were a world apart from the office computers with which IBM executives were familiar. They were not off-the-shelf, mass-produced items but the preserve of a dedicated network of enthusiasts who formed computer clubs, met in user groups, wrote their own newsletters and were keen and able to build their own machinery from scratch.[26]

The most popular of these early machines, the Altair 8800 minicomputer, was sold as a mail-order kit and had first appeared on the front of *Popular Electronics* magazine only a year earlier in 1975. Fully assembled, the Altair consisted of a sheet-metal box with a perplexing array of toggle switches and LEDs on the front. It was an esoteric piece of specialist equipment with relatively little computing power, made purely to learn about the intricacies of computer programming rather than as a readily usable device on which to run productive software. Even when the IBM Yellow Bird was presented to the executive board in 1976, two such home computer enthusiasts, Steve Wozniak and Steve Jobs, had just started selling the Apple 1 computer as a motherboard that hobbyists built into their own handmade wooden cases. The home computer was, at this point, not a beautifully finished commercial

Prototype of the IBM 'Yellow Bird', 1976.

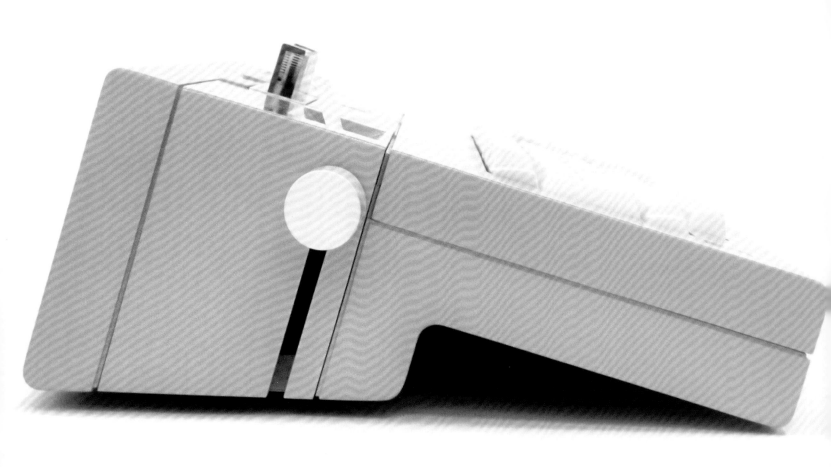
Side view of the IBM 'Yellow Bird' with data cartridge inserted, 1976.

product but often a hastily assembled collection of soldered components and exposed wires. Faced with such products, a cautious response to a potential mass market for personal computers is perhaps more understandable. As Hardy admits, the Yellow Bird was 'kind of ahead of the game'.[27]

The Yellow Bird design proposal might not have changed the mind of the executive board, but it was an exciting enough product concept to galvanize Bill Lowe into continuing on the path of pushing for the personal computer. Despite the continuing setbacks, Lowe had more work in mind for Tom Hardy.

Designer **TOM HARDY**
Date of design work **1977**
Project **IBM AQUARIUS**
Client **IBM**

DESCRIPTION **A self-contained computer in an injection-moulded case with full keyboard, slots for solid-state memory expansion and software cartridges. Display was achieved by connecting the computer to a domestic television. Dimensions 445 by 205 by 100 mm (17.5 by 8 by 4 in).**

The IBM Aquarius was a fully working prototype of a home computer using the very latest in solid-state memory technology. It was particularly advanced for its time and was considered by the team that developed it to be a 'game changer'—something that would have made others follow in its footsteps.

Built in an attempt to demonstrate the potential of the personal computer market to IBM's executive board, the concept was perceived as being too risky to invest in and was turned down. The decision was a disappointment for the team and gave numerous other companies the time to produce their own machines and become established as part of an emerging market.

Studio photograph of the IBM Aquarius in a domestic kitchen setting, 1977.

The IBM Aquarius designed by Tom Hardy, 1977.

The head of IBM's General Systems Division, Bill Lowe, was not deterred by the frequent rebuttals to his proposal that the company should develop a personal computer. By 1977, a number of developments around home computers were taking place elsewhere, and it was becoming more and more clear to Lowe that personal computers were going to be a serious business.

Three home computers were launched in 1977 that, between them, kick-started a whole new industry of associated hardware and third-party software. The first was announced in January, when the Commodore PET 2001 was demonstrated at the Consumer Electronics Show in Chicago. It was a completely self-contained item in a futuristically styled angular case, with 4 kb of memory, a built-in calculator-style keyboard, monochrome 9-in. CRT monitor and cassette drive. Originally offered for sale at $495 in April, advanced orders were so good that by the time it finally shipped in October it cost $795. The second machine, the Apple II (originally the 'Apple]['), was introduced at the West Coast Computer Faire in April and went on sale in June for the comparatively huge sum of $1,298. It was presented in an angular sloped casing with a full keyboard but sold as standard without any monitor or disk drive, just 4 kb of memory and two games paddles. It was, however, the first to have colour graphics. Finally, in August, Tandy announced the TRS-80, priced at $399, or $599 with a 12-in. monochrome monitor and cassette drive, which was delivered to customers in November. Together, these machines presented the home computer not as an arcane, handmade hobbyist artefact but as a professionally finished, off-the-shelf product—these machines were information appliances.

At Lowe's request, Hardy began to work on more serious projects to support the initiative to develop an IBM personal computer. Lowe had decided early in the year that the next product proposal had to be more than a mock-up showing what could be achieved if the time was put into creating a working prototype and actually had to be a working product that the executive board could try out for themselves. Given the developments in the industry, Lowe had managed to get a number of people at IBM interested in working on such a product and had put together a small team of engineers who shared his vision. Tom Hardy was the assigned industrial designer to the project. He recalled, 'The engineers built the working prototype to fit the industrial design model, which was the right way to do it as opposed to having all the hardware lying around as a bench model and having the industrial designer do some fictitious thing that didn't work. They actually got to work on the model, and we called it "Aquarius". I can't recall the reason why—probably New Age, Aquarius and all that thing.'[28]

The objective of this project was to design a personal computer that would be simple for non-specialist users to operate. Like the Yellow Bird designed the previous year, it had the obligatory full-travel IBM keyboard and the red 'paddle' power switch on the top of the casing, and it was designed to be hooked up to a television set to keep the cost down. In a neat touch, the connectors were all placed at the rear of the unit, beneath a removable cover. Once the cables were plugged in, the cover was replaced, leaving the cables to emerge unobtrusively from underneath the computer.

The similarities with the Yellow Bird ended there, though. The IBM Aquarius was designed to be upgradeable and expandable by adding more memory capacity using solid-state 'bubble memory' cartridges, which were inserted into a slot on the left-hand side of the computer's casing. Bubble memory was a relatively recent technology that had been developed at Bell Laboratories in the late 1960s and early 1970s. It was of huge interest to the computing industry because at the time, disk drives were notoriously expensive, complicated to manufacture and prone to failure. Given that bubble memory relied on patterns (or 'bubbles') being created in sheet material by electromagnets, it had no moving parts. The bubbles corresponded to pieces of information, and as the bubbles themselves were extremely small, very high-density data storage could be achieved in a very small space.[29] The industry fully expected bubble memory to replace all forms of core memory, magnetic tapes and disks.[30]

The Aquarius's desired ease of use was achieved through employing other solid-state cartridges, which were inserted into the right-hand side of the computer. Each of these contained one of a range of software packages, including 'word processing, accounting and personal book keeping, games and educational cartridges for kids, and were about half an inch thick'.[31] Each software cartridge came with a corresponding underlay card that was inserted into a slot towards the front of the unit on the right-hand side and, when in place, covered a touch-sensitive panel.

Side view of the IBM Aquarius showing the slots for software cartridges and keypads.

Plan view of the IBM Aquarius with software cartridge and keypad.

Through an array of square apertures in the top of the casing, this card provided a dedicated function keypad to go with each program. Hardy was impressed by the result: 'The engineers packaged all this technology into the case and they made it work.'

Hardy chose the dark-red colour of the casing for similar reasons to the Yellow Bird. 'It was a personal thing, back then everything was neutral, but in big systems we had large coloured panels, and for years IBM had made coloured typewriters for the office, so I thought why not?'[32] As with the IBM Yellow Bird, a series of studio photographs were taken of the Aquarius in domestic settings to help Lowe to explain the computer in context, including one of the computer being used in a kitchen to store recipes.

The IBM Aquarius was evidently way ahead of the competition—it was more powerful and had more memory than other home computers on the market, it was a lot smaller than, for example, the Apple II, and it took advantage of the very latest technology. Hardy was convinced that the IBM Aquarius was a game changer: 'The Aquarius would have blown the socks off of everybody. I felt, and a lot of people felt this was going to be a big deal and make IBM believe in this whole business.' Yet when Lowe demonstrated the prototype to the executive board, he hit a problem. Engineers on the board were wary of bubble memory—it was to some extent an unproven technology, it was occasionally volatile and there were potential risks in investing heavily in something so new. Hardy recalls, 'Because it was relatively new for these kinds of applications and whereas the team had gotten this thing to work, the company just didn't want to take a chance and push it. IBM should have been able to do it. If they would have pushed that technology and put all of their resources behind it like they'd done with other things in the past, a lot of folks thought that it would have been successful and would have just blown the whole thing wide open.'³³ But the board would not be swayed.

The decision was a heavy blow to the team. Hardy says, 'I was so pleased with it. I think it was one of the best projects I ever did when I was working in that company. It was just heart breaking. Some engineers left the company when this wasn't done. One engineer had come over from England to drive this, and he turned round and left, saying "IBM should do it if anyone's going to do it."'

By not entering the market at this point, IBM's competitors—Commodore, Apple and Tandy—had a clearer field, without having to compete with a well-resourced industry giant. As a result, they managed to become leading players in the emerging market for home computers.

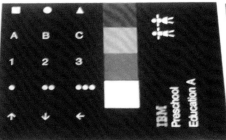

A selection of solid-state software cartridges and function keypads for different IBM Aquarius programs.

Project XEROX NOTETAKER
Client XEROX CORPORATION
Designer ALAN KAY
Date of design work 1978

DESCRIPTION **Portable computer in a carrying case 550 by 345 by 190 mm (21.75 by 13.5 by 7.5 in.) with a 7-in. touch-sensitive screen, floppy disk drive, keyboard and mouse.**

The Xerox Notetaker was a prototype born out of sheer desperation. Its designer, Alan Kay, had for years dreamed of a portable computer that would meet every possible computing need and would be simple enough to operate that a child could use it intuitively. The limitations of technology, though, put that dream in the distant future.

In order to prove his ideas would work, Kay and his research team decided to develop a 'stepping stone' towards this ultimate portable computer. It was to have all the latest available technology currently in use in the advanced computers within Xerox crammed into a single case. The computer itself was achieved, but unfortunately, there was no way of making it light enough to carry around easily. The technology did not come cheap either, and Xerox did not see the product as a viable proposition. Although it never went into production, the Notetaker inspired a whole generation of 'luggable' computers.

Working prototype of the Xerox Notetaker, 1978.

The state-of-the-art GRiD Compass, 1982.

The drive to achieve portability in the computer industry was an incredibly strong one. For years, various people struggled to achieve this within the limits of available technology, but for a long time, the size and weight of internal components meant a fully functional, self-contained portable computer remained a fantasy. Manufacturers were instead limited to producing portable terminals, which would access mainframes in the same way as the remote terminals connected to mainframes that were standard at the time, with access being provided through the telephone network.

Yet, as expected, computer components quickly got smaller as time passed, and eventually the dream of having a portable computer with all the capabilities of a desktop computer became a reality. In the early 1970s, prototype portable machines were produced (see, for example, the IBM SCAMP project), but packaging such functionality small enough meant that certain compromises had to be made, and weight always had to be balanced against cost. To illustrate this, it is interesting to examine two very different computers that were designed and developed at almost exactly the same time. Without compromising on performance or technology, the first laptop computer from GRiD, the 1982 Compass, weighed only 4.2 kg (9.25 lbs) thanks to its use of the latest miniaturized components and a magnesium case.[34] It had 256 kb of memory, 384 kb internal data storage and a 6-in. flat-panel plasma display. Its cost, though, was an incredible $8,150. At the other end of the scale, and often described as the first successful mass-produced portable computer, the Osborne 1 was first produced in 1981 and was described as being 'as portable as a suitcase full of bricks'.[35] The Osborne 1 was, in effect, a very standard, low-performance desktop computer fitted into a case the size of a sewing machine, with a small 5-in. CRT display, 64 kb of memory and two 5.25-in. floppy disk drives, weighing around 12 kg (24 lbs). Consequently, its cost was a far more reasonable $1,795.

The form of the Osborne 1 set the scene for a number of low-cost portable computers that copied its configuration almost exactly, but few people realize that the form was based on a much earlier prototype that in fact had far more computing power than would be seen in the marketplace for many years to come. That machine was the Xerox Notetaker, which was developed by Alan Kay at the Xerox Palo Alto Research Center (PARC) in California.

Kay had been promoting the concept of the ultimate portable computer, the 'Dynabook' (see Xerox Dynabook), since the early

The less capable but much cheaper Osborne 1, 1981.

1970s and was fully aware that it would take time and a huge research and development program to bring it to fruition. In 1976, becoming more and more frustrated as it became apparent that such a research program was unlikely to happen within the foreseeable future, Kay and his Research Learning Group attempted to develop a product that would take advantage of the latest technology and act as a 'stepping stone' to his dream of realizing the Dynabook. Kay, along with team members Adele Goldberg, Larry Tesler and Doug Fairbairn, sketched out a design for a computer that would be able to fit inside a suitcase but have some of the functionality as the Alto computer developed at Xerox in 1972. The Alto was a highly advanced machine and the first to use a mouse and graphical user interface, or GUI (this was the machine that Steve Jobs was shown prior to producing the Apple Lisa and Macintosh computers). Xerox management considered the idea of a portable, microprocessor-based version of Alto as looking 'too far ahead of the curve'.[36] and would not support its development, leaving the group with no option but to develop the machine themselves.

Kay's aim, as with the Dynabook, was to have a computer that could be operated by anyone, irrespective of age or experience. It was to be, in effect, an electronic notebook, and he called it the 'Notetaker'. The team's notion was that children would even use the Notetaker, taking it to school to use during the day and bringing it home with them to do their homework. This gave them

The Xerox Alto computer, 1973.

the target for the finished product in performance and size—it had to be powerful, easy to use and light enough for children to carry.[37]

Over the next two years, Kay and his team strived to pack as much functionality as they could into the Notetaker. The final design consisted of a large carrying case with a removable lid that housed the keyboard and held the mouse. Removing the lid revealed a prototype 7-in. touch-sensitive monochrome display and a floppy disk drive. It used a special version of the Alto 'Smalltalk' GUI, had 128 kb of memory (a large amount for the time) and had an ethernet board so that it could be networked to other computers. It also boasted a microphone, stereo speakers, and a rechargeable battery. The components inside each one were worth around $10,000.[38] The downside to this high specification was the weight—in the end the Notetaker weighed over 20 kg (45 lbs),[39] meaning there was no way it could have been carried by children.

By June 1978,[40] Kay had produced ten fully working prototypes of the Notetaker to prove to Xerox management that it was indeed possible to produce a portable computer with Alto's capabilities. It was tested in the field and even used successfully during airplane flights.[41] Larry Tesler spent the best part of a year presenting the machine to Xerox division executives across the country, but despite promises to take it up, nothing happened. In despair, Kay announced he was leaving to take a sabbatical, but he never returned to PARC.[42] The Notetaker was never put into production, but through connections with other people working in the industry, who were well aware of the Notetaker and built the Osborne 1 and its clones, the design concept of a 'luggable' computer lived on.

Project **IBM 'ATARI' PC**
Client **IBM**
Designer **TOM HARDY**
Date of design work **1979**

DESCRIPTION **A self-contained home computer in an injection-moulded plastic case with full keyboard and slots for joysticks, memory upgrades and software cartridges. Dimensions: 405 by 320 by 115 mm (16 by 12.5 by 4.5 in.).**

The established manufacturers of serious business computers had denied the possibility of a lucrative market for personal computers for a number of years. So when home computers very suddenly took off in 1977, it took the industry by storm. Like other companies, IBM was surprised at the speed of growth of this new market and was understandably keen to join in.

It is not widely known that at the time, there was a very real possibility that IBM could have chosen to buy an existing manufacturer of personal computers and rebrand their products as a direct route into the market. It would certainly have been quicker than developing its own product.

The company being considered as a possible purchase was the video game company Atari, which had recently announced its first range of home computers. An 'IBM' version of one of Atari's computers was fully costed and prototyped, but in the end, the takeover never took place. IBM chose instead to develop its own wildly successful product. But things could have been very different: 'If IBM had bought Atari, this would have been the IBM PC.'[43]

Prototype of the IBM 'Atari' PC designed by Tom Hardy, 1979.

The original Atari 800 Home Computer, designed by Kevin McKinsey, 1978.

Bill Lowe, the lab director for Entry Level Systems in Boca Raton, Florida, part of the General Systems Division of IBM, had been pushing the company's executive board to develop a personal computer for a number of years. In fact he had done so ever since the development of the IBM SCAMP prototype in 1973, but to no avail. No matter how hard he tried, it seems he could not convince those in charge that there was a real market for a personal computer or that the necessary investment was a worthwhile risk.[44] Then, seemingly out of nowhere, the home computer market exploded. The three home computers launched in 1977—Tandy's TRS-80, the Commodore PET and the Apple II—had had an enormous impact. Over the following year, the tidal wave of sales of those home computers was so strong that it influenced many other companies, many of them small start-up companies, to introduce new products. IBM executives saw the potential market at last. Fearing that they were being left behind, they took notice of Lowe's advice and in 1979, after the software programme VisiCalc for the Apple II had made the machine of interest to business users, finally asked him to produce a personal computer.

The only problem with IBM producing a home computer was the sheer complexity of the corporate systems involved. The business model for any new IBM product was tortuous, including putting a team of people and a management/reporting structure in place, going through all the developmental checkpoints and so on. At each stage of the development there were safety issues that had to be dealt with, product testing to be approved and reports to be filed. As a consequence of this cumbersome procedure, the product cycle in the company took a minimum of two to three years.[45] The fact was that the market for personal computers was moving far faster than IBM could. In fact, the convoluted mechanisms involved were well known even outside IBM, to the extent that one analyst was quoted as saying that 'IBM bringing out a personal computer would be like teaching an elephant to tap dance'.[46] Lowe knew the company had to move faster than the procedures allowed and needed a way to get around the corporate red tape involved.

Lowe's strategy was to present the IBM executive board with a 'straw man proposal'—a viable but unpalatable alternative that would force the relevant issues at the highest level. Lowe had some discreet discussions with a computer company that had proved it was able to move quickly: Atari, based in California. As a result of these talks, Lowe asked IBM designer Tom Hardy to go to California, meet a particular person there and only there so it could be kept quiet, and learn all about one of their latest machines, the Atari 800. Hardy recalls, 'So I went out and I met them. I got all the product information and I saw the product, talked to engineers, understood how it was packaged and all that and got a machine and brought it back with me to Florida.'[47]

Atari Inc. was originally started by Nolan Bushnell in 1971 as Syzygy Engineering, the firm that designed and built the first arcade video game—*Computer Space*. The firm became Atari in 1972 and had huge success with *Pong*—their arcade version of the Magnavox Odyssey tennis game.

In 1976, Bushnell began to develop a video game console for home use. Bushnell was convinced of the potential of the machine but also knew that bringing it to market would be extremely expensive. In order to bring in the necessary investment, Bushnell sold Atari to Warner Communications for an estimated $28 million. The Atari video game console was successfully released in 1977—the same year that saw the launch of the three hugely popular home computers. Atari's reaction was

The original Atari 800 Home Computer, with cartridge cover raised, 1978.

much the same as IBM's—the company wanted to get into the clearly lucrative home computer market. Atari hurriedly created a Home Computer Division and immediately started development on their own range of home computers. This was an ironic move, considering that Bushnell had turned down the chance to produce the Apple 1 only a year earlier!

Atari's new line of personal computers was based around a then-powerful 8-bit processor, which enabled an unprecedented level of graphics and sound quality. The machines were announced in late 1978 as the Atari 400 and Atari 800, although they were not widely available until late 1979. The Atari 400 was an entry-level machine aimed at younger users, priced at $550 and having 4 kb of RAM and a 'childproof' membrane keyboard. The upmarket version, and the one of interest to Bill Lowe, was the Atari 800, priced at $1,000 and having 8 kb of RAM and a full standard keyboard.[48]

An Atari in-house designer, Kevin McKinsey, did the industrial design work on the Atari 800 casing. The intention was to give the computer a familiar, friendly look, similar to a standard home typewriter.[49] Although it was a full personal computer, the machine betrayed its gaming roots with the four slots at the front, which were controller sockets for joysticks. The Atari 800 was designed from the outset to be easily expandable up to 48 kb through the use of additional memory cards, which were plugged into the top of the machine behind a lift-up cover in the casing. In front of the memory cards, software cartridges could be inserted. These

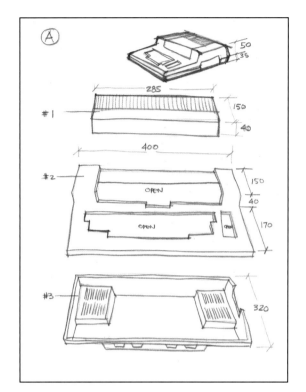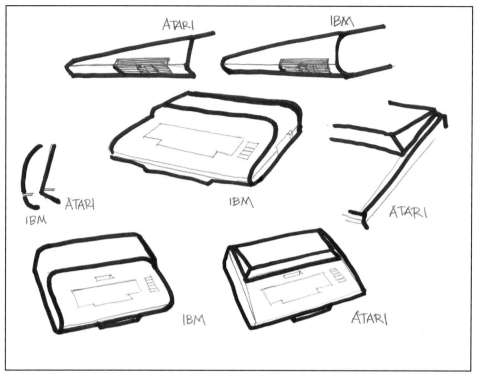

Sketches by Tom Hardy, analyzing the Atari mouldings for costing and considering how to 'convert' the form into an IBM product.

cartridges allowed direct access to various packages, including BASIC programming as well as a range of popular games, which could be run on the machine without the tedious loading of software through compact cassettes as on many competitors' machines. The large metal housing for these cards and cartridges doubled up as the shielding for the computer's motherboard, which because it contained TV circuitry had to be very heavily shielded to prevent interference.

Hardy was given a straightforward brief—reverse-engineer and repackage the Atari 800 and turn it into an IBM personal computer. Some of the Atari components would be retained, but it would be a more serious computer—the IBM added value would appear in the form of a redesigned motherboard, more memory and 'the cherished IBM keyboard with its feel and reliability'.[50] Hardy recalls that his 'mission' was to change the design of it so that it didn't look anything like the Atari. Certain aspects had to be retained, including the four game controller sockets, because games might still be added, and the four function keys, although the colours could of course be changed. 'It was going to be pure white, with the function buttons in a combination of blue and green' instead of the Atari's yellow. This followed IBM's colour coding protocols but also distinguished the machine from the Atari.

Hardy did rough sketches of the mouldings that would have to be replaced so that injection-moulding tooling could be priced up by IBM engineers and also produced sketches of how the form would be changed from an Atari to an IBM aesthetic. 'The main idea of what I wanted to do was to get away from the truncated form [of the Atari] and give it our own identity. I wanted to soften it up and create a front and rear radius, this line that curved round and then curved over the front as well. I wanted this hard edge more personal, that was the idea.'[51] The design alterations were fairly straightforward, to the extent that the prototype was built on top of the Atari hardware and was fully functional.

When the IBM Atari PC design was finished in early 1980, Bill Lowe took the prototype to show to the executive board. He presented the whole scenario—that because of the lengthy IBM internal business model for developing new products, if the company wanted to get into the personal computer market quickly enough, then 'the only way to do it within that time frame was to buy a computer company. And he named Atari to the executives.' This was an incredible proposition for the board. 'I think the fundamental response was "What? We are the largest computer company in the world and we can't develop our own personal computer? We're having to buy a toy?"'[52]

Lowe's response was, 'Okay, here's how we can do our own product. We can do it in one year, and get it to market, if you give us the right to run outside the IBM system and we have the authority of Senior Management. If we want our own IBM PC, here's what you have to do. The Chief Officer, Executive Officer and everyone else in the system at the top of the business has to give us the green light, not to be required to go through this group or that group. So the board said "Come back with a plan". We went back with a detailed plan, they gave the green light and that's how the IBM PC was up in one year—they ran through, over, around everybody!'

Lowe went on to deliver on his promise. As IBM's own archive states, 'Lowe picked a group of 12 strategists who worked around the clock to hammer out a plan for hardware, software, manufacturing setup and sales strategy. It was so well-conceived that the basic strategy remained unaltered throughout the product cycle … In sum, the development team broke all the rules. They went outside the traditional boundaries of product development within IBM. They went to outside vendors for most of the parts, went to outside software developers for the operating system and application software, and acted as an independent business unit. Those tactics enabled them to develop and announce the IBM PC in 12 months—at that time faster than any other hardware product in IBM's history.'[53]

Project **SINCLAIR QL+**
Client **SINCLAIR RESEARCH LTD**
Designer **RICK DICKINSON**
Date of design work **1985-1986**

Prototype of the QL+, 1985.

DESCRIPTION **Desktop PC with keyboard and on-board wafer memory unit and external 'mini-stack' of silicon wafers. Dimensions of main unit: 360 by 85 by 46 mm (14.25 by 3.375 by 0.875 in.).**

As a research-led organization, Sinclair had a history of serious investment in pushing technological boundaries—particularly where breaking those boundaries would result in bringing technologically advanced products to a mass audience at a low price. That approach led to some great advances (if not always financial success) in the areas of pocket electronic calculators, miniature televisions and home computers. It was an approach embedded in the very heart of the company.

Consequently, the mid 1980s at Sinclair Research saw the development of a range of radical products, with designs celebrating new technological advances. Key among these were a range of products utilizing 'Wafer Scale Integration'—a proposed method of producing extremely powerful computers at a much lower cost than traditional methods.

The Sinclair QL, 1984.

Following on from Sinclair's successful forays into the home computer market in the early 1980s in the form of the ZX80, ZX81 and ZX Spectrum, there was significant pressure on the company to produce another groundbreaking computer. By 1983, however, the home computer market was not as buoyant as it had been and high-street stores across the country were already starting to discount unsold stock. A product was needed to take Sinclair Research upmarket, away from the entry-level home computer arena and into the proven market for serious business computers.

The project aiming to provide this product was code-named ZX83, originally a top-end portable computer with twin microdrives for storage. Because of numerous problems in development, the project took longer than expected and the product finally emerged in 1984 as the Sinclair QL. 'QL' stood for 'Quantum Leap' because it was seen as a huge jump in the right direction for Sinclair. It was even advertised with a commercial featuring Sir Clive Sinclair leaping over huge models of his competitors' products. Although it was announced in January (just in time to compete for attention with launch of the Apple Macintosh), working units were not available until much later in the year, causing by now familiar complaints about delays.

Despite the QL being promoted as a revolutionary breakthrough product—a 32-bit personal computer costing far less than its competitors—the rise of the IBM PC and MS-DOS standard meant that the market for any 'non-standard' business computer quickly disappeared. The falling cost of computers at this point also wiped out the QL's price advantage. Following lower than expected sales, production of the QL ceased in early 1985.

Through collaboration with the British manufacturer International Computers Ltd, the technology of the QL went on to form the basis of the ICL One Per Desk, a short-lived executive hybrid computer/telephone/address book/messaging system with program cartridges using microdrive storage technology. Rick Dickinson's design of the QL casing, however, had a longer life. The QL design covered Sinclair's established rubber membrane keyboards with hard keys and the use of higher quality mouldings throughout. The highly recognizable keyboard design used a sophisticated and expensive two-shot moulding process from Sweden. With such an investment in tooling, it was always the intention to use the keyboard in as many other Sinclair computers as possible,[54] which it was—firstly as the basis of the ZX Spectrum+ launched towards the end of 1984 and then the ZX

The Sinclair Spectrum 128, 1985.

Spectrum 128 in late 1985. The QL keyboard was also used in the Pandora prototype laptop (see Pandora Laptop), and it also formed the basis of Dickinson's earliest designs to take advantage of the company's move towards Wafer Scale Integration.

Wafer Scale Integration can best be explained by comparison with the production process of regular integrated circuits or 'silicon chips'. Cylinders of semiconductor material (usually silicon) are sliced into very thin 'wafers' anywhere between 275 microns (0.25 mm) to 775 microns (0.75 mm) thick depending on the diameter of wafer produced (standard sizes commonly being between 100 mm (4 in.) and 300 mm (12 in.)). The resulting wafer is then chemically cleaned before having the required circuitry built onto its surface using a variety of microfabrication techniques. The final etched surface contains multiple copies of individual circuits arranged in a regular grid pattern. The processes used to construct the circuits are not infallible. Flaws in the surface of the wafer are unavoidable, meaning some of the individual circuits will be unusable. In order to identify which, all of the circuits are tested using automated equipment, and the faulty circuits are marked. The wafer is then cut into individual chips, the faulty chips are discarded and the usable chips go on to be 'packaged'—placed on a metal frame and encapsulated in plastic, enabling them to be handled for assembly. This inherently inefficient process accounts for a significant proportion of the cost of a finished silicon chip.

The idea behind Wafer Scale Integration is to do away with this process of testing, marking, cutting and disposing of faulty chips altogether and instead use the complete wafer as a single 'super chip'. As a single very large circuit would almost certainly fail (given the problems stated already), the grid arrangement of small, individual circuits is retained and a series of possible 'rewires' are built into the design. Sinclair aimed to develop a system patented in 1972 by the British inventor Ivor Catt and which in 1975 had been the subject of serious consideration for investigation by the Department of Trade and Industry and the Ministry of Defence, who bought the rights to the Intellectual Property. Catt's system proposed that the individual integrated circuits on a wafer should each include a routing switch. Using a test signal, this switch allows adjacent circuits to be used if the test is positive or the circuit to be bypassed if faulty. If one of the chips on a wafer should fail later in use, the same circuitry allows that chip to be bypassed, which in effect creates a self-repairing high-capacity memory circuit.[55] It was believed that this method would make possible the production of very powerful supercomputers from cheaply mass-produced components.

Sinclair developed wafers in-house, and various possible products were explored to take advantage of Wafer Scale Integration technology. One of the first of these was in the form of an add-on for the Sinclair QL. Problems with the amount of memory needed to run its operating system meant that memory expansion had always been an issue for the QL,[56] and various expansion modules were devised using standard RAM and even a microdrive unit. The QL Wafer expansion module used a 100 mm silicon wafer and had an integral die-cast heat sink and Polaroid back-up battery (as used in the Sinclair Flat-screen pocket TV). The module was fully developed, tooled and manufactured, but as a consequence of the poor sales of the QL, it was not sold.[57]

Other prototype products were developed, including the QL+, a planned successor to the QL computer based around the QL body with an internal wafer under the domed surface at the end, and an external hard drive 'waferstack' unit consisting of a series of finned heat sinks containing individual wafers.

Another waferstack product, the Super QL, was designed in a vertical arrangement to save on desk space. The case had a removable tube running down the back, which was a cable management device to neatly route cables out of the base and away. The tube also acted as a chimney to vent internal heat out of the box. The black fins and blocks on the front of the computer are the external parts of the heat sinks to which the internal wafers are attached. The triangular black block at the bottom is a power supply, positioned there to improve the physical stability of the computer by widening its footprint and adding weight.

The potential of Wafer Scale Integration to allow the economic construction of supercomputers led to other, much more ambitious concepts that unfortunately never got farther than the drawing board. These included designs for a 'Mega PC Waferstack'—a modular, 1 m high tower computer designed for the Sinclair spin-off company Anamartic Ltd. In the end, though, Wafer Scale Integration proved not to be quite as economical as had been hoped. The technology did get to market in other server-based computer components through Anamartic, but at this point, Sinclair was sold to Amstrad, and the potential line of development ceased altogether.

(Facing page) QL Wafer Expansion Module, 1985.

Prototype of the Super QL (rear view), 1986.

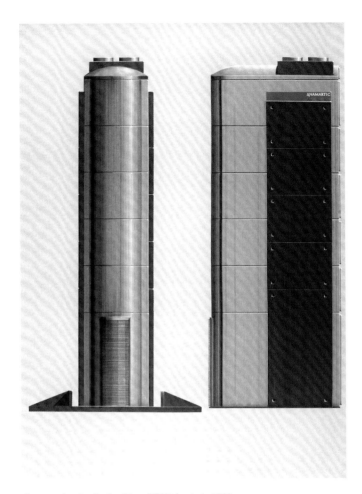

Concept drawing for the Mega PC Waferstack, 1986.

Project **DRAGON PROFESSIONAL**
Client **DRAGON DATA LTD**
Designer **PAT CENTRE, CAMBRIDGE**
Date of design work **1983**

DESCRIPTION **A self-contained personal computer in an injection-moulded casing incorporating a full keyboard and twin 3.5-in. floppy disk drives. Dimensions 380 by 320 by 130 mm (15 by 12.625 by 5.125 in.).**

A home computer made by a toy company and sold through a high-street chemist sounds unlikely as a recipe for success, but that's exactly what happened with the launch of the Dragon 32. It was such a success, in fact, that it caused a chain of financial difficulties from which the company never really recovered and led to its takeover by a major international organization.

In a series of moves that saw the company lose credibility with both their existing and potential new customers, the makers attempted to move into the more lucrative business market with computers such as the Dragon Professional. Despite significant financial investment, the company failed to deliver new products within a reasonable time frame and disappeared before they could reach the market.

The GEC Dragon Professional, 1983.

Packaging of the Dragon 32, 1982.

The form of the casing echoed that of earlier home computers such as the Apple][and that of some of its competitors, in that it was a beige self-contained unit with a wedge form and a keyboard built into the sloping top surface. Mettoy moulded the cases for the computer on the injection-moulding machines used to produce its ranges of toys, with the finished computers being assembled by Dragon Data itself. Full production started in July 1982 and the Dragon 32 was formally released for sale at £199.50 in August. Mettoy already had stockists in place for its existing toys, which included the UK-wide high-street chemists Boots. Boots was persuaded to stock the new computer, bringing it to the attention of a wide consumer base.

Pitched directly against the bestselling American Tandy TRS-80 Color Computer (even using the same microprocessor) but costing £150 less, the Dragon 32 initially received high interest. When shortly following its release Acorn Computers, Commodore and Sinclair Research all had problems supplying their products to retail outlets,[58] demand for the Dragon 32 far outstripped supply and stretched the manufacturing capacity and financial resources of Mettoy to breaking point.

Within the space of three months, it was clear the new division could not continue as it was. Banks were not keen to fund the expansion of Dragon Data, and a number of external investors including the Welsh Development Agency and Pru-tech (the high-tech investment division of Prudential Insurance) had to be persuaded to refinance the company. As a result, Mettoy were left owning just over 15 per cent of its own creation.[59]

The market for home computers in the United Kingdom was extremely buoyant in the early 1980s. The Welsh toy manufacturer Mettoy Ltd decided in late 1981 to enter this market and the company's managing director, Tony Clarke, set up a new division called Dragon Data Ltd. He commissioned the Cambridge technology company PAT (PA Technology) Centre, to build them a new microcomputer and a prototype of their first product, the Dragon, was shown to Mettoy's board in November the same year. PAT Centre was charged with engineering a full production version.

Initially, the Dragon had 16 kb of memory, but following Sinclair's launch of the Spectrum and the announcement that a 48 kb version would soon be made available, the specification was upgraded to 32 kb. This meant that the first production run of 10,000 Dragons had to be hastily upgraded before being shipped.

The financial restructuring allowed the assembly of the Dragon 32 to be contracted out in order to try to meet demand, and as a condition of the deal with the Welsh Development Agency, Dragon Data moved to a new, larger factory outside of the parent company Mettoy. By the following spring, Dragon Data was the largest privately owned company in Wales, and its computer was being stocked by a number of different high-street stores, placing it in direct competition with Sinclair, Commodore and Oric (all stocked by the UK high-street stationers W. H. Smith).[60] Bolstered by the appearance of a huge amount of third-party software (mostly games, but also some basic business and educational titles), the machine sold over 100,000 units in its first year.[61]

Meanwhile, the encouraging start to Dragon Data's trading meant that the PAT Centre had been working on expanding the Dragon range, looking at various floppy disk drives, enhanced

"What would I do with a GEC Dragon 64?"

1. AVOID THUNDERSTORMS. By linking into Prestel, you could call up detailed weather reports at any time of day or night.

2. CHECK THE SPELLING OF EVERY WORD YOU WRITE — INCLUDING THE TECHNICAL ONES. If you're not too sure of your spelling, the Spellcheck program will put you right in seconds.

3. WORK OUT HOW MANY TINS OF CAT-FOOD YOU HAVE IN STOCK. And work out which are the fastest and most profitable lines.

4. CONTACT EVERY ONE OF YOUR CUSTOMERS. Many businesses use GEC Dragon's Mailmerge program to type the same letter, personalised to suit every one of thousands of customers. All you do is write the basic letter, give it the names and addresses, then sit back and wait for the replies.

5. STOP WORLD WAR III BEFORE IT STARTS. Naturally, there are literally hundreds of computer games to while away the extra spare time your GEC Dragon 64 has created for you.

6. FIND A CURE FOR INSOMNIA. Instead of lying awake worrying about the business, get the GEC Dragon 64 to keep it all under control.

7. CHECK THAT EINSTEIN GOT IT RIGHT. When it comes to advanced maths and formulae, the GEC Dragon is little short of a genius.

8. SEND REPORTS OVER THE PHONE. You can send urgent messages or information through Prestel to the GEC Dragon 'Mailbox', for collection by other computer users.

9. SPEND SUNDAY MORNING IN BED. The biggest benefit of them all if you're in business on your own. By taking care of all the details, the GEC Dragon lets you concentrate on the more important things in life.

10. WORK OUT WHAT YOU'LL BE WORTH WHEN YOU RETIRE. Play the investment and insurance companies at their own game and work out EXACTLY how big your nest egg will be when the great day arrives.

11. SPEND AN EVENING WITH NEIL DIAMOND. With a little help from Prestel, you can book seats at almost any show or theatre without even leaving your armchair.

12. LEARN TO FLY A PLANE. We even know someone who has created their own program to simulate the controls of a light aircraft.

13. WRITE A THESIS. If you're not very good at typing, or keep changing your mind, the GEC Dragon word processing program lets you edit, change, add extra pieces and delete. Then when your masterpiece is finally ready to type, just press a button and sit back.

14. CLAIM YOUR FORTUNE ON THE POOLS. The GEC Dragon 64 also gives you immediate access to a mass of sports information available through Prestel.

15. BOOK YOUR HOLIDAYS. Check the availability of practically any holiday you care to think of. Then make a reservation on the one you like best.

But that's just for starters. Later, we'll show you lots more ways the GEC Dragon 64 can make life simpler.

You can buy the GEC Dragon computer and a wide range of accessories and software from the better computer shops, major stores and GEC dealers.

It's proof that, now GEC and Dragon have got together, we're really going to start turning it on for the small business and serious computer user.

And to whet your appetite still further, we've produced a 12-page colour brochure that tells you how to get the most out of a GEC Dragon 64. It's called 'Your passport to professional software'. It's yours free in exchange for the coupon below.

GEC DRAGON COMPUTERS

To: GEC Dragon Customer Services, Tripsgate House, Gladstone Drive, Staple Hill, Bristol BS16 4RU.
Please send me a copy of Your Passport to Professional Software.

Name
Address
Postcode

Or if you would like information on the rest of our products – please tick the appropriate box
☐ Dragon 32 ☐ Dragon 64 ☐ Dragon Accessories

Advertisement for the GEC-badged Dragon 64 in Dragon User magazine, July 1984.

The GEC Dragon Professional as featured in Personal Computer World, *August 1984.*

graphics capabilities and disk operating systems. Two products were planned to move Dragon Data upmarket—the first (Project Alpha) to compete directly with the BBC Micro, and the second a 128 kb machine (Project Beta) to compete with IBM's recent launch of the IBM PC.[62] At one point, after the directors had been impressed by Apple's Lisa, there was even talk of developing a computer with a graphical user interface.[63]

In order to progress expansion before the longer-term projects could deliver, the company meanwhile launched its next product, the Dragon 64. This fairly straightforward memory upgrade (the product was visually identical apart from a change of the case to a blue/grey colour) was launched in August 1983 in the United States (where it was manufactured by Tano Corporation of New Orleans)[64] and September 1983 in the United Kingdom, and it was met with enthusiasm on both sides of the Atlantic.

Unfortunately for Dragon Data, September 1983 was something of a watershed. Tony Clarke's forecasts predicting that the level of Dragon sales experienced in the run-up to Christmas 1982 would continue throughout 1983 turned out not to be correct.

John Linney, a software engineer at Dragon at the time, reflects, 'It was a straightforward question of supply exceeding demand. After a very successful Christmas we continued to build product at a prodigious rate, but didn't anticipate the dramatic drop-off in demand. I think back then, the seasonality of home computer sales as a novelty item were not well understood.'[65] The resulting lower income over the summer, when the firm needed more resources to ramp up production for the following Christmas, caused serious cash-flow problems. At this point, the UK consumer and defence electronics conglomerate General Electric Company (GEC) became involved.[66] A rescue package in the form of £2.5 million of guaranteed loans was organized, and GEC executive Brian Moore replaced Tony Clarke as managing director of Dragon Data Ltd.[67] Mettoy itself was less fortunate and went into receivership at the end of October 1983.[68]

Moore continued the company's aims of moving into the business arena and backed the continuation of Project Alpha and Project Beta, as well as the creation of a new multitasking operating system, OS-9; although Moore was not keen on discussing future projects in case competitors found out too much or in case it took sales away from the existing product lines.[69] In John Linney's view, targeting the business market was seen as a way of combating the seasonal sales of the home computer. Entering the market with a design that could be quickly produced using existing and plentiful parts would solve the problem. Yet moving to a business audience also meant changes in the marketing of the company—away from its long-standing arrangements with high-street retailers such as Boots the Chemist and towards new relationships with specialist retailers. Consequently, it was not long before the Dragon 32 and 64 were being sold off in the high street at bargain basement prices. As if this was not damaging enough, Dragon Data's attempt to change the public's perception of the company meant that by March 1984, sales and marketing were handled by a GEC subsidiary and the products were rebranded as 'GEC Dragon'. The computing press soon displayed double-page-spread advertisements with the new brand, explaining how their computers were so much more versatile than being used just to play games, although the visual appearance of the advertisements was not overly businesslike. In fact, these moves formed 'a series of PR and marketing hashes that destroyed most of the company's credibility as a manufacturer of home computers'.[70]

Despite Moore's protestations against premature announcements, the result of Project Alpha—the GEC Dragon Professional—was announced in May 1984. It was a transportable 64 kb machine with a built-in modem and integral 3.5-in. Sony disk drives and was priced at £700 (with a single disk drive) or £850 (with twin disk drives). A small number of early prototypes were made for internal, technical assessment and a handful of prototypes for reviewers using 'production' mouldings.[71] The casing looked similar to their existing products (even using the same base moulding) with the new disk drives placed above the keyboard on the new top moulding, forming a more angular wedge. The computer had no cooling fan and suffered problems with overheating from the power supply. Reviewers noted it ran well for a while before starting to act strangely and then cut out after an hour's use. Attempts to solve this were hastily made by drilling a circle of small holes in the fabricated casings of the first prototypes and the addition of a series of vents in the top of the production moulding, but the unit still ran too hot for comfort.

A prototype badged as the GEC Dragon Professional was shown at the Consumer Electronics Trade Exhibition at the end of May, only days before Dragon Data went into receivership. The editorial in the July 1984 issue of *Dragon User* magazine that featured an article on the new computer responded to the 'sad news', stating that 'the company has confidence in its new products and will be using its best endeavours in helping the receivers to explore ways of continuing trading'. A Spanish company, Eurohard SA, stepped in to take over production of the Dragon 32 and 64, but it struggled to maintain interest in what was by then fast becoming an outdated piece of technology.

Another prototype of the new business machine was sent to *Personal Computer World* for review just before the company's demise, and a review was actually published in the August 1984 issue despite the obvious unavailability of the computer at that time. It concluded, 'If Dragon recommences trading … If the quoted prices are realistic … If the cooling problems are overcome and … If a new incarnation of Dragon can regain lost credibility … then even counting the [Sinclair] QL the Dragon Professional would be hard to beat for small businesses.'

Sadly, despite rumoured interest from GEC, Tandy and Philips,[72] Dragon Data never did recommence trading.

Pen Computing

The field of pen computing, which hit its peak in the early 1990s, was perhaps the most prolific in producing items of vapourware. The mere idea of such a product was enough to generate huge amounts of interest from the industry and financial backing from venture capitalists. For a few short years it was constantly being reported that every major manufacturer had a pen computer under development.[1] In most cases, that is exactly where they stayed. The few pen-based machines that did make it to market had relatively little success, with most disappearing quietly soon after their launch.

Pen computers—computers that are operated by writing commands on the screen rather than typing commands into a keyboard—were seen as the 'logical' next step forward in the development of personal and portable computers. Typing is not a 'natural' process in the same way that writing is, and it was thought that a computer that took advantage of such a natural process as writing would have a significant advantage over its competitors and be the product everybody would want to own—making the successful company the market leader. Judging by the coverage of the subject in the computing industry press, there was massive potential for such a machine. However, despite decades of research and development and colossal financial investments from all concerned, it proved much harder than thought to properly solve the technical difficulties involved.

Using a pen as a device to operate a computer has a far longer history than many people realize. Shortly after the Second World War, the 'Whirlwind' computer built at Massachusetts Institute of Technology (MIT) used 'light pens' to interact with a computer monitor. That machine formed the basis of a whole series of computers using the technology, including the largest computers ever built, the SAGE system (Semi-Automatic Ground Environment) for air defence, and the TX-2 computer, which was used to develop the first ever computer drawing package, Sketchpad, in 1963, in which a light pen drew lines directly onto the computer display.

Making a computer understand written commands, though, required a lot more technological development. The computer needed to be able to translate the movement of the pen into an understandable command. In other words, it needed to be able to recognize handwriting. Early attempts to solve this problem were undertaken at the Advanced Research Projects Agency (ARPA) in the early 1960s. By 1964, the research team had produced the RAND Tablet, or Grafacon—a drawing surface containing hundreds of sensors that could locate the special stylus writing on it to within one-hundredth of an inch. Alongside this, they developed software called GRAphic Input Language (GRAIL) that allowed programming a computer to be achieved by writing text and drawing flowcharts on the RAND Tablet. GRAIL formed the basis of word processing without a keyboard that is taken for granted today.

Finally, to enable the commands to be written directly onto a screen rather than a separate dedicated writing surface required the development of a transparent touch-sensitive screen. This emerged out of work initially done in the early 1970s to digitize data from strip chart recorders, which required an X-Y location on a printed-paper graph to be measured. This was achieved using conductive sheets sandwiched between insulating layers of paper, which were then pierced by a metallic pin, sending an electrical signal to a voltmeter. The company that did this work, Elograph, had discussions with its patent lawyer about possibly developing a transparent version for use with computers, and Siemens provided the necessary research funding.

Once all of these technologies were in place, pen computers could finally be developed. The first such machine to be produced was the Linus Write-Top in 1987. This consisted of a touch-sensitive LCD screen and tethered stylus, which could be clipped onto the processing unit it was attached to by cable to make a portable pen computer. The various technologies were at a very early stage, however, and the functionality of the computer was limited. Linus Technologies closed its doors in 1990.

More successful than the Write-Top was the 1989 GRiDPad. This was a completely self-contained unit with the processing unit in the same slim case as a large LCD screen. The case had five function buttons, and a tethered stylus was used to write on the screen. In some strictly defined markets it was a very functional product. It was mostly used by various field agents and representatives in the insurance and medical industries, who had to regularly fill out large numbers of simple forms. As a hugely glorified electronic clipboard, early pen computers worked well, but there remained a reluctance to use such products in a broader context. One company, Momenta, learned to its cost that aiming such a device directly at the executive market was no guarantee of success. Despite having a whole host of advanced features aimed directly at making an executive's life easier and the publicity generated from appearing on over twenty magazine covers on its launch in 1991, the company lasted less than one year. As one critic put it, 'Momenta was a monumental flop.'[2]

Purely technical reasons were not the only issue, although the man behind the GRiDPad, Jeff Hawkins, says that the inescapable feel of writing on glass with a stylus was indeed a major drawback.[3] In truth, the market for the computer just wasn't there. It is likely that more senior executives resented being seen writing on something resembling a clipboard as much as they resented being seen typing on computers that looked like electronic typewriters before the commercialization of the graphic interface and mouse made the office computer the ubiquitous machine it is today. Whatever the reason, just a few years after the idea of pen computing was first mooted, it moved from being heralded as the future of computing to becoming the laughing stock of the industry. In an article describing the state of the industry in 1994, the editor of *Pen Computing Magazine* wrote, 'To say that the pen computing industry was struggling was a vast understatement. "Dying", "reviled", "ridiculed" would more aptly describe it.'[4]

Computer companies didn't give up easily, though, and they kept on promising new, ever smaller, more powerful and more functional pen computers. A full-sized pen computer seemed set never to hit the market. Pen computers changed direction, morphing into personal digital assistants (PDAs) and becoming more focused handheld products with a different functionality. Even then, take-up was initially very slow. Products such as Apple's Newton MessagePad of 1993 received massive publicity, but it too was subject to various functionality problems, with poor handwriting recognition being cited as the main reason for its poor sales. Eventually, Jeff Hawkins's new company Palm solved the usability issues and developed the Palm Pilot, and PDAs really took off. True pen computing was still heavily promoted, not least by Microsoft.[5] Various forms of pen computers were tried, including convertible laptops and tablets, but it was not until the advent of screens that could be easily operated by the fingers instead of a stylus that people started to become interested. Along came Apple, with their revolutionary multi-touch, gesture-based screen used on the iPhone, which gave such products an element of kudos and cachet of cool and convinced the market that their 2010 tablet computer, the iPad, was the way to go.

Project: **XEROX DYNABOOK**
Client: **XEROX CORPORATION**
Designer: **ALAN KAY**
Date of design work: **1968**

DESCRIPTION **Notebook-sized 300 by 225 mm by 6.35 mm (12 by 9 by 0.25 in.) tablet computer with stylus and keyboard operation, removable data storage and wireless networking capability.**

The Dynabook was a conceptual product designed by the maverick computer scientist Alan Kay in 1968. The technology to produce it was nowhere near available, and its target audience was one the industry had never considered. It was as radical as anything proposed in science fiction stories, a fact which Kay himself admitted. But it was technically feasible given the investment of enough money, time and effort. Kay's reputation as a brilliant pioneer meant many took the proposal very seriously. As such, the Dynabook inspired numerous attempts to produce computers that were smaller, highly functional and easier to operate. Within Xerox, though, attitudes to Kay's work were not as positive.

Kay's prediction of what could be achieved and the form it should take proved to be incredibly accurate, and forty years later, many commentators have noted the similarities between the Dynabook and the Apple iPad. If only they had listened.

Cardboard mock-up of the Dynabook, 1968.

Imagine having your own self-contained knowledge manipulator in a portable package the size and shape of an ordinary notebook. Suppose it had enough power to outrace your senses of sight and hearing, enough capacity to store for later retrieval thousands of page-equivalents of reference materials, poems, letters, recipes, records, drawings, animations, musical scores, waveforms, dynamic simulations, and anything else you would like to remember and change. We envision a device as small and portable as possible which could both take in and give out information in quantities approaching that of human sensory systems.[6]

This is a description of the Dynabook—the vision of future computing Alan Kay first presented to the computer industry in the early 1970s. It was an incredible piece of foresight, and how he got to that point is a fascinating story.

After working as a computer programmer in the US Air Force, Kay enrolled at the University of Utah in 1966. There, Kay was exposed to a variety of influences that over the course of a few years coalesced into the concept of the Dynabook. The first of these was the PhD thesis by Ivan Sutherland describing his development of 'Sketchpad', the first ever computer graphics program that was fundamental to the later development of pen computing. This was followed by the learning of 'Simula', an early programming language that involved the use of simple mechanisms to control complex processes, which had a huge effect on Kay in developing new ways to structure computation.[7]

In 1967, Kay collaborated with Ed Cheadle, an engineer in a local aerospace company, who wanted to develop a computer that could be used by non-professionals. This became the FLEX machine. Kay's aim was to create a cut-down version of the programming language for Sketchpad or Simula (a FLexible EXtendable language[8]) that could run on a small machine and be operated via a visual display. In the same year, Kay heard a lecture by Doug Engelbart, the inventor of the computer mouse, who visited Utah to talk about his vision of computers as an 'augmentation of human intellect'. Engelbart described his NLS (oN Line System) that he was to publicly demonstrate so famously in 1968. He talked about hypertext linking, graphics, real-time editing and cross-network interactive collaboration. Kay adopted many of these ideas into the FLEX machine.

Kay was well aware of 'Moore's Law': the 1965 prediction that computing power would double every eighteen months. After hearing Engelbart speak, Kay suddenly realized the full impact of Moore's Law. Eventually, the huge mainframe computers he was familiar with would become desktop machines. The world of computing would change from an institutionally based activity involving a small number of corporately controlled machines to an individually based activity involving millions of personal machines. In the future, everybody would own a computer. Consequently, the way a computer was programmed and operated *had* to change.[9]

The next big influence on Kay came from hearing Marvin Minsky lecture about education. Minsky, the founder of the field of artificial intelligence, was speaking out against traditional educational methods and saying that schools were really bad places to teach children how to think about complex situations.[10] It was Minsky who introduced Kay to the educational concepts of Seymour Papert and inspired him to think about the benefits of computing for children.

In 1968, Kay attended the ARPA grad students meeting, where he presented the FLEX machine. While there, he visited Donald Bitzer's lab at the University of Illinois, where he saw Bitzer's prototype of a plasma-panel flat-screen display. Although it was only 1-in. square and only lit up a few pixels, Kay found it 'galvanizing'.[11] He spent the remainder of the conference using Moore's Law to figure out how long it would take to get the transistors in the FLEX machine to fit onto the back of a plasma screen the size of a sheet of notepaper. He very accurately estimated it would be around ten years off.

Later the same year, Kay went to visit other researchers who were working with non-computer professionals. One memorable visit was to RAND in Santa Monica, where he saw GRAIL, the early tablet-based handwriting recognition system. 'It was beautiful', he wrote. 'I realized that the FLEX interface was all wrong.'[12] The other memorable visit was to a school in Lexington, where he met Seymour Papert and watched him use his programming language 'LOGO' to teach twelve-year-olds how to program computers. This, to Kay, was 'a transformative experience'.[13]

During his flight back to Utah, Kay put all the pieces of the jigsaw together and saw what the future of personal computing really should be: everything pointed to the personal computer being a personal dynamic medium—not a tool learned later in life but something learned as a child. On the plane, he drew a cartoon of what such a machine would look like—a tablet computer being used by children for every aspect of their work and play.

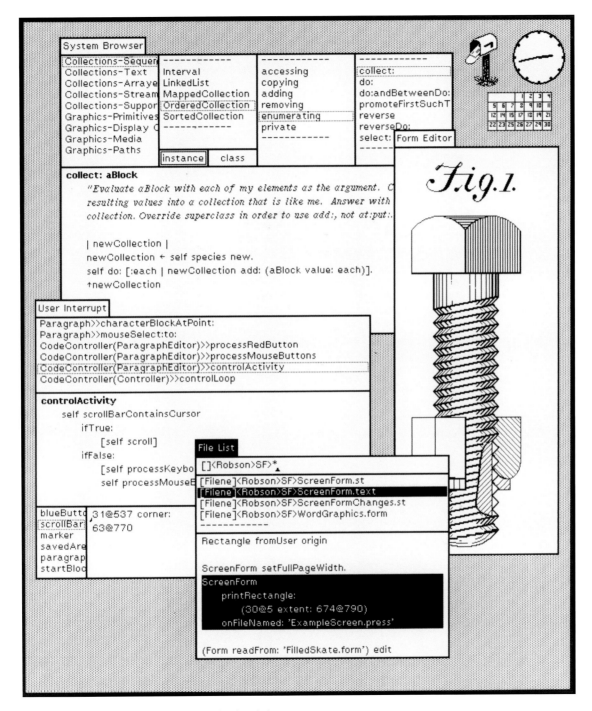

The 'Smalltalk' Graphical User Interface with overlapping windows.

The 'interim Dynabook system' being operated by children, 1973.

He figured from its intended use that the computer should be the size of a notebook and have a friendly but very powerful interface. It should have removable storage and the ability to be wirelessly networked to other computers. Although it should have a stylus to draw sketches on the screen, he added a keyboard, as even if handwriting recognition was perfect, it was faster to type than write. When he got back to Utah, he made a hollow cardboard model of the Dynabook and filled it with lead pellets to determine what weight it should be (which turned out to be less than 4 lbs).[14]

Kay quickly finished his work on FLEX and gained his PhD in 1969.[15] He then worked on artificial intelligence (AI) at Stanford but spent more time thinking about what he started to call 'KiddiKomputers' than AI. When Xerox set up Palo Alto Research Center (PARC), he saw this as an opportunity to develop a KiddiKomp that could be made in quantity and used in experiments to develop the user interface for the Dynabook. The following year, Xerox hired researchers from Berkeley Computer Corporation and Doug Engelbart's Augmentation Research Center, and the people who could work with Kay to 'invent the future'[16] were finally together under one roof. He gathered together the researchers keenest to see the Dynabook project to completion and formed his 'Learning Research Group'.

The group worked to refine the KiddiKomp idea into what was called the miniCOM, which had a bitmap graphic display, a touch-sensitive keyboard, tablet, stylus and mouse and used a simple programming language that became 'Smalltalk' (as in 'programming computers should be a matter of smalltalk'). Kay also famously solved the problem of putting more information on a single display screen by realizing that the windows he had developed for the FLEX machine 'could be made to appear as overlapping documents on a desktop'.[17] That was the birth of the desktop metaphor used on almost every computer today.

Despite the considerable advances made towards realizing Kay's dream, further development was made difficult. New people joined Xerox at the executive level, and not all were interested in supporting Kay's efforts. The company's attention turned more towards development in time-sharing computer systems for the office, Xerox's home territory. Refused money to make the miniCOM, Kay continued to promote his altruistic vision for the future of computing as widely as possible, both within Xerox and to a public audience through published conference papers.[18]

An opportunity to develop the Dynabook idea within Xerox came towards the end of 1972 when in the absence of the executive most strongly opposed to Dynabook, Kay was asked by his colleagues Butler Lampson and Chuck Thacker to join forces (and budgets) and produce a machine that would be a huge step in the right direction—what Kay termed 'an interim Dynabook'. The fifteen prototypes made were hugely successful in experiments in a local school with child users, quickly enabling them to develop their own computer programs, including drawing, painting, animation, games and even circuit board design packages.[19] The interim Dynabook became the Xerox Alto—a computer so advanced everyone within Xerox (including the executive so opposed to the Dynabook) wanted one. Over 2,000 were eventually made, mainly for internal use, but in the end, and for the most perverse of reasons, Xerox still refused to put it into full production because it would mean sales targets of existing product lines would not be met.[20]

By the end of 1975, business imperatives meant that Xerox was inexorably heading back towards the office user and away from the educational needs of children. The Dynabook concept seemed doomed to failure, and Kay tried to revitalize the morale of his Learning Research Group by beginning work on what looked to be a more achievable end product—the Notetaker (see Xerox Notetaker)—but even this proved to be more difficult than imagined.

Xerox never produced the Dynabook, but Kay's concept had a significant lasting impact, mostly because it was conceived not in secret as a corporate profit-making exercise but as a serious and widely disseminated consideration of what computing could achieve for the benefit of society. As one article stated, 'The Dynabook concept has the potential to affect in a very basic way a great number of our society's institutions. The impact of the next new medium, Dynabook, could be earthshaking.'[21] Despite the difficulties he encountered at Xerox, Kay continues to be involved in altruistic work to use low-cost computing tools to educate children to this day. As president of the Viewpoints Research Institute and still working with Seymour Papert, he is deeply involved in Nicholas Negroponte's One Laptop Per Child initiative that 'seeks to create a Dynabook-like "$100 laptop" for every child in the world (especially in the 3rd world)'.[22]

Project **APPLE FIGARO**
Client **APPLE COMPUTER INC.**
Designer **ITALDESIGN, GIUGIARO DESIGN DIVISION**
Date of design work **1991**

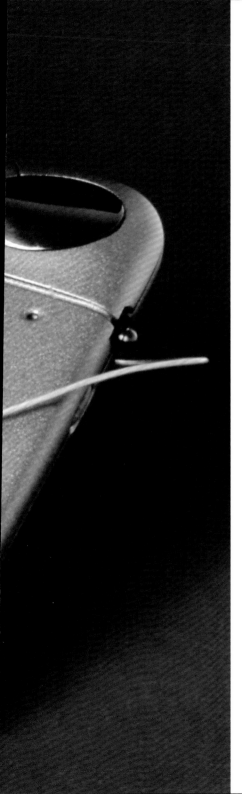

DESCRIPTION **A flat tablet computer with an injection-moulded case, LCD touch-sensitive screen, tethered stylus and built-in infrared transceiver for transferring information between devices. Dimensions: 23 by 30.4 by 3.8 cm (9 by 12 by 1.5 in.).**

One of the earlier attempts by a mainstream computer manufacturer to explore the potential of pen computing was Apple's 'Figaro' project. It aimed to realize, as far as possible, a futuristic vision of computing described by Apple's chief executive officer, John Sculley.

In order to gather a wide range of product ideas, a number of high-profile design consultants were invited to submit their designs for a pen-operated tablet computer as entries to a competition, with extremely varied results. The chosen design consultants, Giugiaro Design, developed the product to a pre-production prototype, but internal politics at Apple, coupled with some uncertainty as to the readiness of the market for such a product, led to the project being cancelled.

'MontBlanc' concept for the Apple Figaro project by Giugiaro Design, 1991.

The roots of the Apple Figaro project stretch back to May 1987 when the then CEO, John Sculley, described his vision of the future of computing devices in his book *Odyssey: Pepsi to Apple: A Journey of Adventure, Ideas, and the Future*. In the book, he describes a device he termed the 'Knowledge Navigator', a portable, voice-activated computer that could search the Internet for information on behalf of its owner. A detailed account of the resulting project appears in Paul Kunkel's book, *Apple Design: The Work of the Apple Industrial Design Group*. According to Kunkel, Apple's in-house design team produced a concept model of the Knowledge Navigator, which looked like a book that could fold in half, had no keyboard and included a camera for videoconferencing and a built-in handle to carry the device around. It was an inspiring concept, and Apple human interface designer Sue Booker was given the project, code-named 'Figaro', to explore the possibility of turning the idea as much as possible into reality. Figaro was intended from the start to be a high-end product, utilizing the latest technology without compromise. The product specification had a high target price of $6,000, a touch-sensitive screen and a pen-based interface, a miniature disk drive and an infrared transceiver to allow information to be beamed between devices. By May 1989, all the design parameters of shape, size and functionality had been resolved, but no work had been done on the industrial design, so nobody had any idea what Figaro might look like.[23]

In parallel to Sue Booker's work on Figaro, steps were being taken to change the overall direction of Apple's designs. Hartmut Esslinger and his consultancy, frog design, had served Apple well since the early 1980s, when he began to work on a variety of products, culminating in the creation of the 'Snow White' design language to be applied to all Apple products. The Snow White

First Figaro concept by Smart Design, 1989.

First Figaro concept by Smart Design in transit, 1989.

Second Figaro concept by Smart Design, 1989.

language of simple forms with minimal texture and crisp detailing, shallow ventilation and decorative grooves, no draft angles, small radii and a very pale grey colour was highly successful, putting design at the top of Apple's agenda. But, as with all things, the designs eventually started to look dated. Richard Jordan, a Stanford graduate who had joined Apple in 1978, became manager of the Industrial Design and Product Groups in 1985 and almost immediately started to phase out Esslinger's Snow White language and begin a new, worldwide search for Apple's next 'design superstar'. After a long and fruitless search, Jordan met with the world-famous Giorgetto Giugiaro in May 1989 in his design studio in Turin and felt he had achieved his goal. Giugiaro was asked to create a design language that would unite but differentiate between high-end, mid-range and portable products and to produce a few key designs to serve as examples to Apple's in-house teams that would implement the design language.

At the same time, Sue Booker was looking for a way to achieve a new, unique product identity for Figaro and felt it could not be met by Apple's in-house design team. She instead commissioned the services of external design consultants to competitively create different designs. The designers were Ettore Sottsass from Milan,

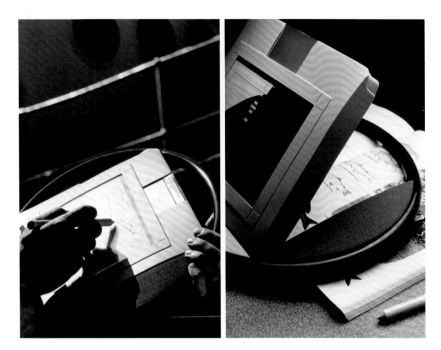

Second Figaro concept by Smart Design in situ, 1989.

Smart Design from New York, and Doug Patton from Los Angeles. To these, Booker added Giugiaro. Tom Dair of Smart Design remembers the brief for the Figaro project as being to produce two non-working appearance models and five images of each design to show how they would be used, all within thirty days.[24] 'It was a fairly open, single page brief, mostly taken up with a schematic showing a block form factor, and indication of a screen, an estimated weight and mention of a battery. They clearly wanted something with a tablet form, but little else was specified.'[25]

The finished results of the Figaro design competition were presented in September 1989. Sottsass's designs were very simple. Patton's were beautifully detailed and considered, but too typically 'Silicon Valley high-tech'. In comparison, Smart Design's work was more radical. Their first concept was a sharp-edged, rectangular aluminium box forming the 'pages' of a book-like object wrapped in a soft, brown leather cover. The edges of the aluminium block were pierced with an array of small holes for ventilation and to act as a speaker cover. It had a slot to accept a large memory card on which were to be stored programs and data, a built-in camera and the large screen to be written on with a curved 'bone-like' stylus. Smart Design's second concept was a square grey panel with a touch-sensitive screen housed inside a blue oval hoop that doubled as a handle. The design encouraged the product to be used in portrait as well as landscape mode. On one side, the gap between the central body and the hoop was filled with a loudspeaker covered in fabric, and the special stylus, which resembled a yellow crayon, had a pointed end for entering data and a rounded pink end shaped like an eraser to delete information. The body could also be pivoted out of the hoop, which then acted as a stand to allow the device to be used on a desktop and enabled it to scan documents placed underneath.[26]

In complete contrast to Smart Design's concepts, Giugiaro had produced three very straightforward-looking slate-like designs, one all white, one grey and one which was black with a red stripe and had a curved edge that allowed the user to hold it comfortably. It looked to Booker as if this could have been designed for one of Giugiaro's other clients, Nikon, as it bore a strong resemblance to his design of the Nikon F4 camera.[27] However, after an internal review, it was decided that Giugiaro's work was the most suitable direction for Figaro. In December 1989, Giugiaro was asked to develop his concepts further, and he delivered four more proposals in February 1990. These were softer, more colourful blue/green and grey designs with the preferred option having a tethered stylus, angled slits for a speaker cover and a large red 'eye' in the top right-hand corner as a cover for the infrared transceiver. Although the designs were not well liked by Apple's in-house design team, who saw them as 'too cute' or simply 'not Apple enough',[28] the user response from focus group testing was very positive.

In early 1990, the internal politics at Apple were coming to a head, and a number of changes occurred that had a direct impact on the Figaro project. Richard Jordan finally managed to persuade one of his favourite designers, Robert Brunner, to join Apple as director of the Industrial Design Group, although Brunner admitted he had concerns over Giugiaro's role. The head of research and development, Jean-Louis Gassée, resigned in April after a falling-out with CEO John Sculley, quickly followed by Steve Sakoman, a computer engineer who was responsible for overseeing the Figaro line of product development. Sculley brought in marketing

First Figaro concepts from Giugiaro Design, 1989.

First Figaro concepts from Giugiaro Design, 1989.

Revised Apple Figaro concepts by Giugiaro Design, 1990.

Apple Figaro 'red eye' concept by Giugiaro Design, 1990.

manager Michael Tchao, who immediately split the Figaro project into three parts: a high-end large notebook computer (essentially Figaro) called the 'Newton Plus', a middle-sized version called 'Newton' and a smaller, pocket-sized version called 'Pocket Newt'. Nobody was interested in the middle-sized version, leaving the Newton Plus and the Pocket Newt to be taken forward as potential new products.

Tchao, like Robert Brunner, was not a fan of Giugiaro's designs, despite them proving most popular with potential users. He spoke to Sculley and reached an agreement that Brunner be asked to produce designs for the Newton Plus to be tested against the concepts from Giugiaro. Rising to the challenge, Brunner, along with an old consultancy colleague, Ken Wood, produced a range of concepts, but despite their appeal to the in-house design team, focus group testing still pointed to Giugiaro's designs. The 'red eye' concept was developed further into a refined, silver/grey version of the Newton Plus called 'MontBlanc' in February 1991 that was fully engineered and ready for production. The Newton team was unsure whether to push the fully developed Newton Plus or the untested Pocket Newt when further consumer testing showed that MontBlanc's key feature, the large infrared transceiver, was unlikely to be of much interest to users. It was also seen as too big and too expensive. Tchao persuaded Sculley to put the Newton Plus on hold and to divert the research and development effort into the Pocket Newt instead. At this point, Sue Booker left Apple rather than see her Newton Plus interface trimmed down for a small product.

A slightly sinister-looking version of the Pocket Newt nicknamed 'Batman', with a fold-over cover inspired by the hood of a Corvette, was developed between November 1991 and February 1992. The Newton Plus project was finally cancelled in May 1992, and resources were put into a productionized version of Batman called 'Junior'. The Junior became Apple's first PDA, the Newton MessagePad 100, launched in August 1993.

The Apple Newton MessagePad 100, 1993.

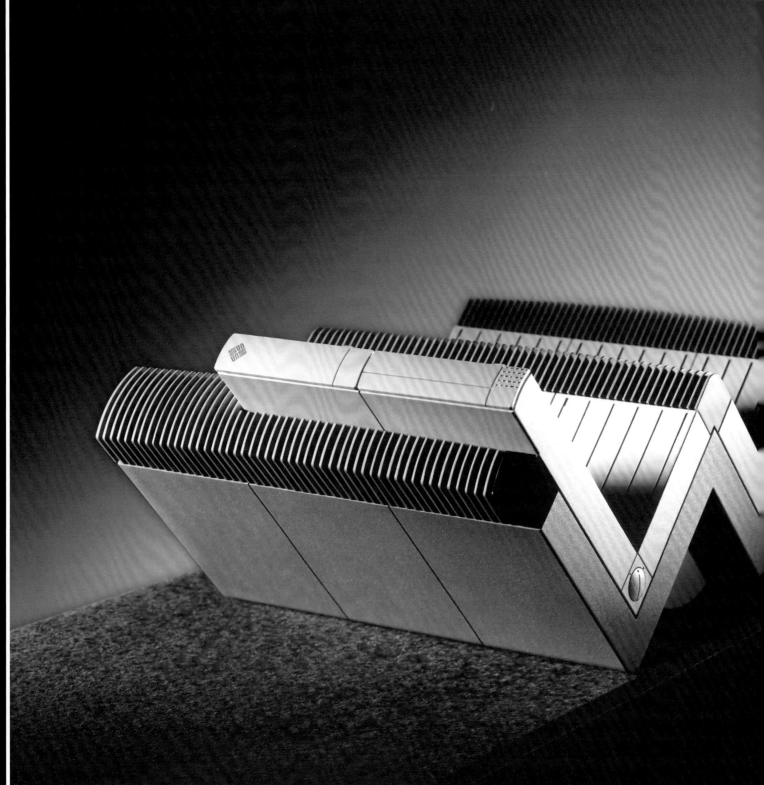

Project **SUN MODULAR COMPUTER** Designer **PAUL BRADLEY AND JOCHEN BACKS, MATRIX PRODUCT DESIGN**
Client **SUN MICROSYSTEMS** Date of design work **1990**

DESCRIPTION **An injection-moulded desktop central processing unit with hard drive and disk drive and a casing consisting of a series of expansion units edged with metal finned heat sinks.**

Sun Microsystems was a forward-looking manufacturer of high-end computers and a high-profile proponent of open computer systems. They were a major player in the development of open-source software, including the cross-platform 'Java' programming language and the free OpenOffice suite of business productivity software.[29]

At a time when the company was developing a whole series of radical cutting-edge products, one particularly radical concept did not receive the backing required to go into production. Matrix Product Design's concepts housed a modular, pen-driven computer system, based on reduced instruction set computing (RISC), developed by one of Sun's internal R&D teams, which was seen as a step too far even for them.

Sun modular computer central processing unit concept, 1990.

The first SUN machine started as a computer project by a Stanford University graduate student, Andreas Bechtolsheim, and was built out of spare parts he sourced from the computer science department and local suppliers.[30] At the time, an emerging interest of the computer industry was in the area of workstations. Workstations, like personal computers, were designed to be used by a single person but were expensive, specialist pieces of fast computer hardware with significantly more processing power and higher-resolution graphics, allowing them to be used for running complex design and engineering software such as 3D CAD modelling and rendering and Finite Element Analysis. Bechtolsheim was interested in building a workstation that could be easily networked for a communications project with which he was involved.

Bechtolsheim's approach to the design of his workstation was in sharp contrast to the established order. Instead of keeping strict control over the end product by using custom-made hardware and a proprietary operating system, he used off-the-shelf components to make the computer more affordable and AT&T's UNIX operating system, which, as a popular choice of other manufacturers, would enable the easy sharing of data with other companies' workstations and make the SUN workstation easier to network. 'SUN' was an acronym for Stanford University Network, and from 1981, Bechtolsheim started to sell licences for the computer at $10,000 each.[31]

Shortly afterwards Bechtolsheim was joined by three more Stanford graduates, Vinod Khosla, Scott McNealy and Bill Joy, and the company Sun Microsystems, Inc., was founded in February 1982. The decision to use standard hardware components and operating system software was in many ways similar to IBM's choices made around the IBM PC and, similarly, initially proved to be a very successful strategy. It meant that Sun was able to quickly enter the market with a competitively priced product, but as with the IBM PC, it also meant that the design could easily be copied. However, sales of the first production models, the Sun-1 and Sun-2 workstations, were so strong that the company was soon a very profitable one. Initially, the majority of sales were to universities, but by the end of the decade, Sun Microsystems had become a force to be reckoned with throughout the mainstream computer industry.

As a forward-looking, technology-led company, Sun was interested in where technology was headed, and the company's R&D teams were constantly exploring new directions. One such direction, based around RISC architecture, was for a modular computer system. RISC stands for reduced instruction set computing, which put very simply is a way of using very simple computer commands that allow a computer to operate much more quickly. The modular hardware the R&D team had developed enabled a basic entry-level computer system to be very flexibly expanded into ever more complex systems, even up to something close to a supercomputer.[32] The system worked so well and had so much potential that the R&D team was keen to promote the idea to Sun management in an inspiring way. They commissioned the Palo Alto–based Matrix Product Design consultancy to develop concepts for products that could incorporate the technology to best advantage and be presented to management alongside the proof of concept technical hardware.

Matrix design director Paul Bradley, along with designer Jochen Backs, carried out the industrial design work on the project. Between them, they produced a range of design concepts addressing the brief. From these, two particular designs were chosen and taken forward to more resolved design proposals that were fully hard-modelled. The first design was for a zigzag-form central processing unit that could be oriented in different ways, being laid flat on a desk surface or placed vertically to reduce its footprint, making it more like a mini tower unit. The basic unit had a single RISC processor chip, a hard drive and a selection of floppy disk drives as required, with a number of expansion modules slotting into the body of the zigzag form to increase the computer's capacity. Each of these modules contained a RISC chip mounted onto a block with metal fins that were needed to act as heat sinks to keep the temperature within operating parameters. Backs remembers that 'the heat dissipation of the very hot modular micro chips was a big issue and a challenge—how to express it aesthetically in an interesting and new way. We wanted to highlight the technology through showcasing the exposed heat sinks.'[33] In an entry-level system, these modules would be 'blanks' that could be replaced by working modules as required for expansion. The central processing unit itself could be connected via cables to peripheral devices including monitors, printers, plotters and so on to form a flexible computer system.

The second design chosen for further development was for a more self-contained pen computer system consisting of a central processing unit in the form of a vertical metal frame. Mounted

(Facing page) Sun modular computer central processing unit concept in upright position.

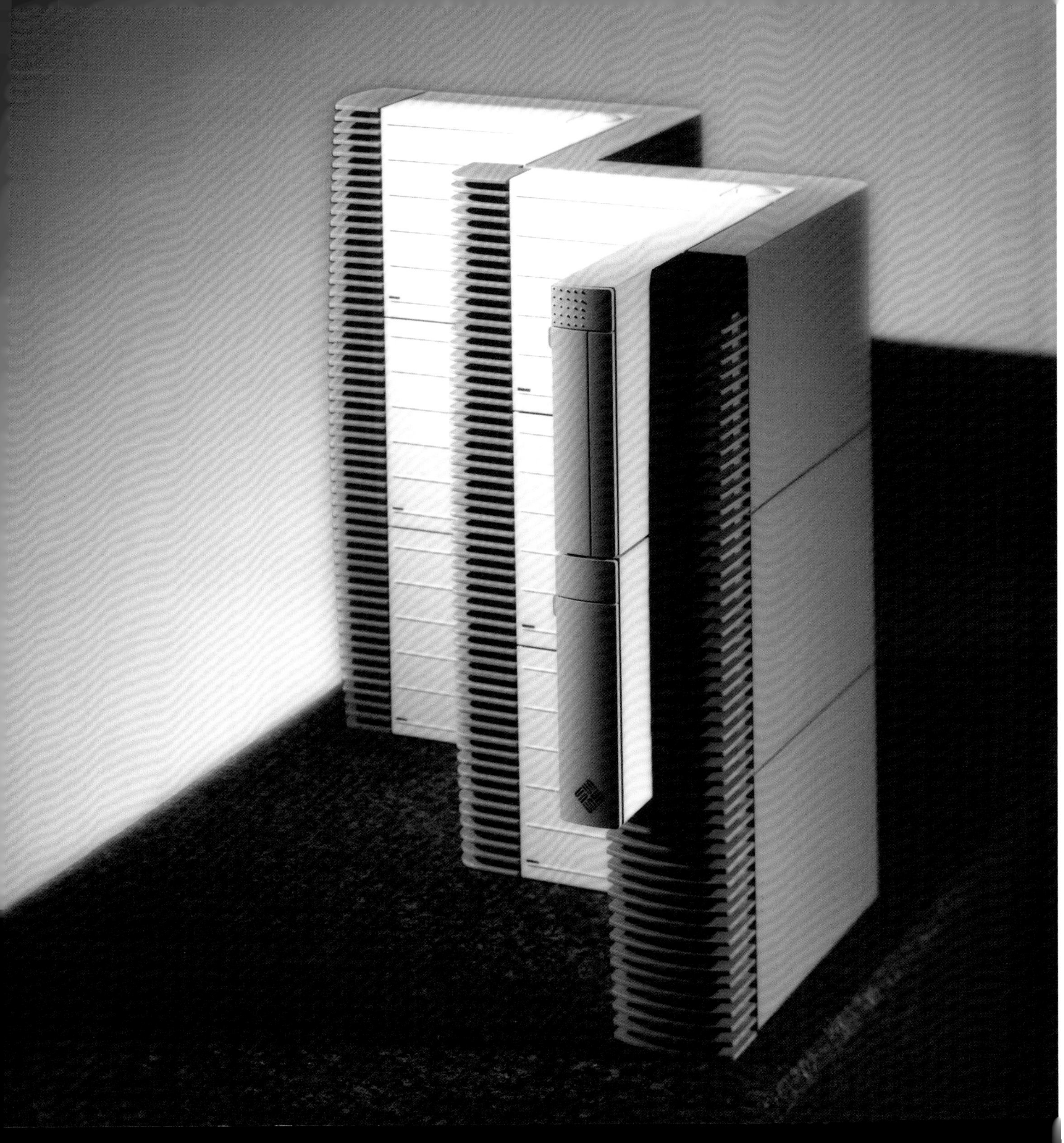

onto this frame was a touch-sensitive display panel that could be written on with a dedicated smart stylus housed in the base of the product. The display panel could be angled away from the central frame and had a built-in camera at its top centre. The central vertical frame itself could be pivoted out to create a plate into which the pale-grey expansion modules containing RISC chips could be slotted. This made the modules completely surrounded by the finned metal plate, which acted as a huge heat sink. The base unit contained disk drives hidden behind curved cover plates.

Although all concerned put a lot of effort into the project, Sun management did not take up the concept for a modular RISC-based computer. It is possible that internal decisions were made to follow a different architecture design or that the massive investment it would have taken to put the concept into full production was considered too big a risk. Although considered futuristic, design consultancies such as Matrix were working at the time on a number of exploratory concepts for different manufacturers, all bringing touch screens into the desktop. Touch screens were the sort of elements, along with features such as built-in cameras, that 'didn't really appear until a long time later, but at the time, the tech community thought it was going to happen'.[34]

Although this particular line of development was not taken up, Sun continued to expand rapidly as a company, taking over a whole raft of other computer hardware, software and server and network system companies. At the end of the 1990s, Sun was part of the frenzy of investment in computer businesses now referred to as the 'Dot Com Bubble' and became massively more valuable as a company. As a consequence, Sun invested heavily in staff, office buildings, manufacturing plants and capital equipment on a global scale, but when the bubble burst in 2000, the sales of what were still relatively expensive, high-end machines suffered a sharp decline. As a result, many staff were made redundant, a number of factories were closed and the value of the company fell dramatically. Although the company briefly returned to profitability in the mid 2000s, because of poor management decisions,[35] Sun quickly fell back into debt. On resigning his position, the then CEO of the company, Jonathan Schwartz, notably delivered the news of his departure to staff and shareholders through a tweet in the form of a haiku verse: 'Financial crisis / Stalled too many customers / CEO no more.'[36] After a protracted battle with IBM, Sun Microsystems was finally sold to the database management systems company Oracle Corporation in 2009.

(Facing page) Self-contained modular pen computer concept, 1990.

Project **GO PENPOINT COMPUTER**
Client **GO CORPORATION**
Designer **PAUL BRADLEY, MATRIX DESIGN**
Date of design work **1988–1991**

DESCRIPTION **Injection-moulded casing resembling a notebook, with a large LCD screen. Dimensions: 228 by 330 by 19 mm (9 by 13 by 0.75 in.). The device has a fold-over cover and an untethered stylus stored inside the casing.**

The GO PenPoint Computer was the result of 'the modern scientific version of religious epiphany'.[37] On a flight to San Francisco in February 1987, the computer scientist Jerry Kaplan had a lengthy discussion with Mitchell Kapor (one of the founders of Lotus Development Corporation). Juggling with phones, organizers, chargers, adapters and a large portable Compaq 286 computer in the restricted space of his private jet, Kapor started to type the scribblings from various scraps of paper into his computer, stating, 'I wish there was some way for me to get all this stuff directly into the computer and skip the paper.'[38]

The two businessmen discussed how they could combine the latest technological developments to create a much smaller, lighter computer that could be written on rather than typed into—a computer that acted like a notepad rather than a typewriter. Such a device, they figured, could solve all of the information-handling problems encountered by travelling executives. Turning the epiphany into reality proved to be harder than expected.

Prototype of the GO PenPoint Computer with external accessories, 1991.

Jerry Kaplan wrote a whole book, *Start Up: A Silicon Valley Adventure*, about the GO computer.[39] In a story of high-finance, high-pressure business with enough intrigue, double-dealing and treachery to fill a John Le Carré novel, Kaplan described his efforts to bring a pen computer to the market and, in doing so, to change the face of personal computing.

According to Kaplan, shortly after being asked by Mitchell Kapor to run the start-up company they named 'GO Corporation', he quickly put a team of core specialists together. The aim was to design a computer that consisted of mainly a large LCD screen, which would be written on with a special stylus containing an electronic sensor. Through an electromagnetic grid over the surface of the LCD, the computer would be able to identify the location of the sensor in the stylus and so make the corresponding part of the LCD react accordingly—electronic ink. The interface was to reflect a notebook, with tabbed 'pages' that could be flicked through to see different documents. To handle the hardware side Kaplan recruited engineers Kevin Doren and Celeste Baranski, while on the software side he engaged the services of Robert Carr and Baranski's husband, Mike Ouye. An initial public offering in 1987 raised the necessary capital, and Kaplan promised the investors the delivery of a deskbound prototype pen computer by June 1988.

However, by early 1988, it was clear that delivering a product was not going to be plain sailing. 'It seemed as though we'd uncover a new and more dreadful problem everyday,' Kaplan wrote, describing a list of catastrophes. With the hardware, they found that flat-panel LCD displays couldn't take the pressure of being written on with a pen and that using a protective sheet of glass not only made the tip of the pen appear to float over the electronic ink but made the act of writing feel like 'roller skating on ice'. Attempts to remedy this by etching the glass surface meant that the already hazy image on the LCD became even more blurred. The software needed for a completely new pen-based operating system provided other obstacles. Although the circuitry was based on Intel's fastest 286 processor, there were unacceptable delays between writing on the screen and the ink appearing. Each unit also had to be individually calibrated so that the sensor in the pen was in alignment with the screen. When the components warmed up, the alignment would drift, meaning the unit had no idea exactly where the pen was, and even when it did, Kaplan admits, the handwriting recognition was 'awful'.[40]

Solving enough of the problems to put together a deskbound prototype for a successful demonstration in July 1988, Kaplan promised investors a portable version by the following June.[41] Around the same time, GO showed the product to Microsoft founder Bill Gates under a non-disclosure agreement to protect its intellectual property. After initially agreeing to develop the software jointly, Microsoft became convinced it could produce a better product using Windows.[42] In fact, it was far worse than this: according to internal Microsoft sources, Bill Gates saw the GO interface as a serious threat to Windows and set up an internal product development group with a mission to do two things: create their own pen-based software and to 'kill GO Corp'.[43] The well-resourced group set about buying in expertise from start-up companies working in the area of handwriting recognition.

Oblivious to Microsoft's efforts, GO continued on its way to get a real product to market. The first version of the GO Computer was code-named 'Lombard'.[44] Paul Bradley of Matrix Design did the industrial design work, with mechanical engineering aspects carried out by David Kelly Design. To go hand in hand with the operating system, the design deliberately emphasized the 'notebook' metaphor. It had a fold-over leather cover that fitted into the 'spine' along the long edge of the casing to protect the screen and could be folded underneath the unit. The cover was held closed by magnets, but these interfered with the electromagnetic grid over the screen if they came too close together.[45] The stylus took the form of a pen that was stored inside the top corner of the casing, just as a normal pen is inserted into the binding of a traditional spiral-bound notebook. Apart from the large screen, the front of the unit only had an on/off button, providing an unprecedented level of simplicity that one reviewer said made the product look 'much like a concept car'.[46] There was no internal disk drive in the computer, just 8 Mb of static RAM, and a rechargeable NiCad battery supplied the portable power, but the battery life 'wasn't great'.[47] Bradley also designed a series of modular external peripherals, including a disk drive and an external power supply, which clipped together into a single, neat unit, although these were never fully prototyped.

Spending $250,000 per month on product development at this point, by late 1988 the company was fast running out of money. It had a desperate need to find a corporate partner who could support the product development, which was still suffering setbacks. There were problems with the chip that controlled the display and problems with the Japanese supplier of the critically

(Facing page) Prototype of the GO PenPoint Computer and PenPoint Interface, 1991.

PenPoint Gestures	
Selection	
[	Bracket left, selects the word to its left.
]	Bracket right, selects the word to its right.
✓	Tick, displays options for associated objects, icons or tools.
•	Tap, like a mouse click, once to select, twice to open.
•↓	Tap and press, allows text to be dragged around the screen.
Scrolling	
←	Flick left, turn to next page.
→	Flick right, turn to previous page.
↑	Flick up, scroll page down.
↓	Flick down, scroll page up.
Insertion and deletion	
⌴	Insert space, horizontal line indicates the length of the space.
⁀	Pigtail, deletes a character.
↓	Press, deletes a character.
○	Circle, opens an edit pad to modify a selected area.
∧	Caret, inserts text at a point or an entry into a table of contents.
X	Cross out, deletes selected text.
∿	Scratch out, delete selected text.

Examples of the gestures used to enter commands in the PenPoint interface.

important pen sensor. GO had enough parts to build two or three working prototypes, but the first they managed to get working promptly burst into flames when tested. Scavenging undamaged parts to put a second prototype together, a working portable unit was finally achieved on 20 June 1989.[48]

Kaplan presented the prototype to State Farm Mutual Automobile Insurance in a competitive bid for its business against three giants of the industry—IBM, Hewlett-Packard and Wang. State Farm liked the GO product but realized the small company was unlikely to be able to resource full production on its own. State Farm asked GO to enter into partnership with one of its regular suppliers. GO chose to work with IBM and offered a quarter of the company in return for a share of future licensing revenues. In an attempt to beat Microsoft to the market, IBM agreed.[49]

By early 1990, the product development process was once more in trouble, with part of the delay apparently because of IBM constantly changing its requirements of the product. GO still needed further funding, so Kaplan made a press announcement to

One of the sixty-five demonstration machines on the cover of Personal Computer World, April 1991.

generate interest in the company and publicly declare the support of IBM.[50] Interestingly, at this point, GO was announcing not a product but a 'technology' and a 'partnership'—perhaps realizing even then that a finished product was farther away than it cared to admit. The announcement in July 1990 made front-page news in most of the computer publications and did indeed generate interest. Offers of further funding flew in: $5 million from State Farm, $5 million from Intel and others. By now, GO's commercial value was well over $75 million. Apple Computers almost bought GO's technology for its Newton PDA project but in the end developed its own.

Rumours of further delays in the engineering schedule meant that GO was rapidly gaining a reputation as 'a purveyor of "vaporware"'.[51] Yet despite the danger in making public announcements before products were finished, it was forced to make another. Because GO knew Microsoft was going to announce pen-based extensions to Windows, it hurried to launch a development-only version of its operating system called 'PenPoint' in January 1991. A total of sixty-five of the Lombard units, described as 'demonstration machines', were produced for the launch event so that developers could try out the new system, but it was the software, with its tagline of 'The Pen is the Point', that was very much the focus of attention.[52]

Unbeknown to GO, seated in the audience with a video camera was Microsoft employee Wink Thorne. Thorne flew back to Microsoft and the entire pen computing group spent a month 'making sure they could demo everything the GO system had promised'.[53] Although GO had spent $25 million trying to convince the industry that pen computers were the way to go, its thunder was stolen by Microsoft's announcement that it intended to launch a competing pen-based operating system. Only six weeks after the GO launch event, Microsoft was able to demonstrate a system that seemingly did everything GO's did. Kaplan was convinced that Microsoft had stolen their ideas and had been working on them since the first meeting between GO and Bill Gates. What he didn't know was that the Microsoft demo was a fake. Microsoft didn't have any working software at all but had knocked up a demonstration that made it look as if they did. 'It was total bull, pure smoke and mirrors, the apotheosis of vaporware.'[54] Smoke and mirrors it may have been, but it had the desired effect. Many developers decided to hang on and see if they could work with Microsoft and their enormous installed base of established Windows users rather than develop completely new applications for PenPoint.[55]

Microsoft's announcement boosted the industry's fascination with pen computers. Taking advantage of the frenzy, GO distributed the demonstration machines to various journalists hoping to promote interest in its software. The machines appeared on the front cover of a number of widely read computer magazines with bold headlines stating 'Tomorrow's Laptops'[56] and 'All Systems Go!',[57] furthering the mistaken belief that a real, all-new pen computer was just about to appear. The articles inside the magazines were as effusive about pen computing as ever, explaining in great detail the ease of use in writing rather than typing and the use of gestures such as tapping or flicking the pen across the screen as shorthand for commands.

Following the 1991 presentation, efforts continued to produce a full production version of the GO Computer. This project, code-named 'Hyde', was based on Intel's brand-new 386 processor. Although working prototypes were made, this version didn't even reach pre-production. GO Corporation not only was constantly trying to outpace Microsoft but was also coming under increasing pressure from IBM to stop competing with that company in the hardware business.[58] A solution to the software problem was to redevelop PenPoint to run on an even higher specification microprocessor that would outperform the chip that Microsoft was tied into. A solution to the hardware problem was to concentrate solely on developing software and let someone else worry about making the machines on which it would run. Consequently, the hardware side of the company was spun off into a new venture called 'EO', which was later bought by AT&T (See EO Magni Personal Communicator).

In the end, despite incredible amounts of publicity, huge public interest, wide support from the industry and unprecedented levels of investment, the GO Computer never appeared on the market. As Kaplan wrote in his book, 'The real question is not why the project died, but rather why it survived as long as it did with no meaningful sales.'[59]

Project **IBM LEAPFROG 1/2 JET COMPUTER**
Designer **RICHARD SAPPER AND SAM LUCENTE**
Date of design work **1989-1992**
Client **IBM**

DESCRIPTION **Pen-based tablet computer, 270 by 350 by 24 mm (10.5 by 14 by 1 in.) with 10.4-in. colour LCD screen and cordless pen. Detachable base with 'pop-out' keyboard. Upper case in carbon-fibre-reinforced plastic, lower case in magnesium alloy, base unit in vacuum-cast urethane, ABS keyboard.**

Leapfrog was designed as 'a concept car', a forward-looking, high-profile statement of intent, aiming to restore the image of IBM as an innovator of market-leading computer products. It pushed available technology to its limits and promised a new era of design implementation at IBM. With a unique, 'illusory' form inspired by a ream of paper being pushed over, the inch-thick tablet managed to look impossibly thin.

At the time of its production as a working prototype in 1992, the Leapfrog was a well-kept secret shown only to industry insiders. After its public 'debut' in 1993, it was presumed to be a full production item and was written about in design magazines as a commercial product. But it was only ever intended to show that IBM, who once dominated the computer industry, was in no way ready to be written off.

Working prototype of the IBM Leapfrog Tablet Computer, 1992.

The lead designer of the Leapfrog, the Milan-based German designer Richard Sapper, was first hired as a consultant by IBM in 1980 and acted in a supervisory capacity over the company's fifteen design centres across the world.[60] IBM's decision to bring in such a well-known designer to stir up design activity within the company was no unprecedented move. The company had a long tradition of engaging the services of high-profile designers to work alongside and inspire their in-house design teams. As far back as 1956, IBM chairman Thomas Watson Jr. hired the renowned architect and designer Eliot Noyes as 'Consultant Director of Design'—a position he held until his death in 1977. Over the years, other 'name' designers followed, among them the graphic designer Paul Rand and the architect and designer Charles Eames. Throughout the 1950s, 1960s and 1970s, the use of external consultants led to IBM becoming recognized worldwide as an innovative, design-led company.

But things change. When in-house industrial designer Tom Hardy took up the position of corporate head of the IBM Design Program, IBM was aware that despite having massive research and design facilities, it 'was no longer perceived as an innovator'[61] in what had become a highly competitive industry. Traditionally, IBM had followed a system-oriented strategy in which 'products of marked individuality were not welcomed'.[62] To try and counter this, Hardy and Sapper promoted a strategy of 'design differentiation' with the aim being to 'improve IBM's image as an innovator and offering products that exceeded expectations'.[63] Together, they convinced IBM executives that design could be used to 'target individual business users'[64] and persuaded them to provide seed money for new developments. The aim was to design the computing equivalent of 'a "concept car", a sexy computer that would showcase some of the company's design, research and engineering talent in a highly visible "image" product'.[65]

Project 'Leapfrog' was initiated in 1989 as a 'vision for the future, and to get a jump on advanced technology packaging … Leapfrog was to be a catalyst for innovation'.[66] After Hardy and Sapper had secured funding, design product manager Samuel Lucente worked with Sapper on a detailed form proposal. The shape of this hard model was so unusual that the model maker initially misunderstood the drawings as an axonometric drawing of a straightforward box.[67] It was a three-dimensional parallelogram that had initially started out as a series of flat panels laid on top of each other, with each panel moved across and upwards to produce stepped surfaces along the sides. These stepped sides eventually became smooth planar surfaces. The most accurate description, and the way Sapper himself described the 'formal genesis off the idea',[68] is to take a ream of A4 paper and push the stack upwards diagonally so that each sheet of paper slides over the top of the sheet below.

Sapper's inspiration was a traditional inclined reading stand. While working on an earlier 'design differentiation' product, a pen-based tablet sold as the ThinkPad,[69] he was frustrated by the thickness of the product, which made it difficult to write on when it lay on a flat surface. He said,

> The form of the Leapfrog springs from the ideal posture of the hand on a surface, for writing or painting, in short, from the inclination of a reading stand. In a very preliminary sketch I envisaged a sequence of steps that were subsequently changed gradually into a tilted plane. I am particularly attracted by vanishing outlines of this sort.[70]

IBM executives liked the design, and the interdisciplinary team of fifteen designers and engineers were challenged to produce a working prototype that had to fit within Sapper's 1-in.-thick casing. That process took two years.[71] The team, which was dispersed around the globe, communicated almost daily by video link. Research was undertaken to identify the emerging technologies that would be mainstream for personal computers in the mid 1990s. That, in fact, was the inspiration for the project's code name of Leapfrog—the idea of looking down the road four or five years and visualizing what new products future technology would bring.[72] Once identified, all the individual components had to be made small enough and thin enough before design of the final product could begin. Never before had a fully functional computer with the same capabilities as a desktop machine been packaged in such a confined space.[73] Even the thickness of the casing was subject to scrutiny. In order to make it as thin as possible (0.9 mm), it was produced out of carbon-fibre-reinforced plastic—the same material used in the stealth bomber. From the project's research and development stages, IBM gained a number of patents for processes that could be used on other products.[74] Once all the components were finalized, it took only a further six months to produce a batch of a dozen working prototypes that were 'almost identical in every way to the initial concept model developed by Richard Sapper three years earlier'.[75] These twelve prototypes were shown privately to a select group of industry analysts at the industry show Comdex, held in Arizona in November 1992.[76]

I.D

The International Design Magazine
May June 1993 $7

Sex: Why Madonna, Calvin Klein and Harper's Bazaar can't get enough of Fabien Baron

Cars: Five interiors that overcome woodgrain

Can design save IBM?
A concept computer taps the unmined technologies of Big Blue

Cover of I.D. magazine, May–June 1993.

In the final design, the tablet could be connected in either portrait or landscape mode to a base unit christened the 'lilypad' and once attached could be tilted as required. The base unit allowed the computer to connect to external disk drives and local area networks, stored the cordless pen and also contained a full keyboard, which could be stowed away when not required. The idea for this came from previous projects, where people had accidentally knocked keys with their elbows when the tablet was being written on.[77] When removed from the base, the tablet could be written on while carried or supported at an angle by foldaway legs.

The design was well received in the design world, even being exhibited in the Museum of Modern Art in San Francisco. To further promote the successful project, Tom Hardy pushed design magazines to write articles about it, where possible insisting on front-page coverage.[78] In one such article, *I.D.* magazine wrote,

> 'Leapfrog' offers IBM's vision of the next paradigm of office computers—'interpersonal' machines. It can be used at a desk with a keyboard or taken to meetings and used with a pen to take notes and make presentations. While the competition is offering up Personal Digital Assistants, IBM is standing by the office-bound corporate user, creating the successor to the desktop PC.[79]

Despite its obvious appeal to a design-conscious audience, Leapfrog never reached the buying public. Hardy says, 'The prototype cost an enormous amount, but productionizing it would have been way too expensive.'[80] It did, however, fulfil its intended function. Shortly after its public debut, Hardy wrote,

> Working prototypes have been demonstrated at international trade shows, achieved major design honors, and influenced products in both tactical and strategic product development. Like ThinkPad, Leapfrog has been instrumental in helping to rebuild IBM's image as an innovator.[81]

The IBM Leapfrog Tablet Computer on its 'lilypad' base unit.

Project **EO MAGNI PERSONAL COMMUNICATOR** Designer **NAOTO FUKASAWA/PAUL BRADLEY**
Client **EO CORPORATION** Date of design work **1993–1994**

DESCRIPTION **Injection-moulded casing containing a large LCD screen, wireless communication device and a voice-activated computer.**

The EO Magni was part of a range of exploratory concepts commissioned by EO management to assess the market for the next generation of their products to follow the widely lauded but small selling EO 440 Personal Communicator.

EO was heavily funded by the telecommunications giant AT&T. Naturally, given AT&T's position, the company was extremely interested in the convergence of the portable computer and the telephone, which they had broached with the EO 440 in 1993. The next generation of products was intended to explore the possibilities arising from AT&T's belief that in the near future, this convergence would go farther and the role that voice control would play in personal computing would by far surpass that of the written command.

In the end, the numerous delays in getting actual products to market proved too much, and AT&T decided to cease all funding, bringing an end to EO.

Model of the AT&T EO Magni Personal Communicator, 1993–1994.

frogdesign. *global creative network*

frogfignewton

*Design strategy, company name, logo, corporate identity, packaging, collateral and industrial design for **EO, Inc.***

EO Inc. was formed in July 1991 as a spin-off from GO Corporation, which had changed its focus from manufacturing hardware to concentrate on the software development of its PenPoint operating system. The engineering team of Celeste Baranski, her husband, Mike Ouye, and GO's vice president of operations Paul Hammel formed the nucleus of the new venture.

Interest in pen computing was still high and the number of companies entering a rapidly crowded market was growing. In June 1991, NCR announced the launch of its PenPoint-based machine, the NCR 3125, although NCR admitted it was six to nine months away. To further muddy the already murky waters, GO's supposed partner in pen computing, IBM, announced it would be releasing its own pen-based operating system, Pen OS/2. Then, at the Consumer Electronics Show in Chicago in May 1992, John Sculley, the CEO of Apple Computers, announced the Apple Newton and coined the term 'personal digital assistant' (PDA). Using only a wooden block model and a mocked-up demo of the interface, Sculley got the audience excited about a product that was 'at least a year off'.[82] Suddenly, the pressure was really on EO to get a product onto the market.

AT&T was keen to finance EO to achieve this and funded it with 'tens of millions of dollars'.[83] As a communications company that had previously been barred from entering the computer industry on monopoly grounds, AT&T had good reason to explore combining computer and telephone technology. The company even made a series of promotional advertisements with the tagline 'You Will', showcasing its vision for future portable communication devices long before it had products ready for market. The engineering team began two projects: an entry-level machine code-named Thor (which became the EO 440) and a higher-specification version code-named Odin (which became the EO 880). These products were a significant departure from the GO Computer, having a number of extra communications functions.

The industrial design work was carried out by one of San Francisco's leading design firms, frog design, under the direction of Hartmut Esslinger, the designer responsible for Apple's 'Snow White' design language in the 1980s. The large LCD screen was housed in an injection-moulded casing with distinctive semi-circular 'ears' on either side which housed a microphone and

(Facing page) Frog design's advertisement featuring the EO Personal Communicator, Design *magazine, June 1993.*

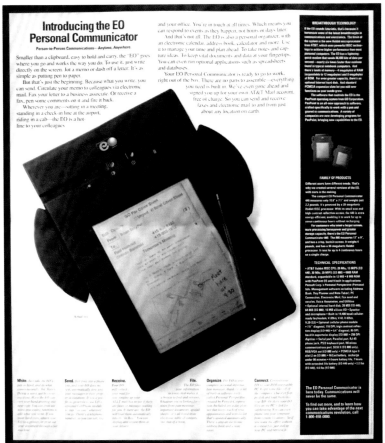

EO promotional brochure, 1992.

speaker. An optional cell phone/modem module with a full-sized telephone handset could be plugged into the top of the unit. Much larger than true PDAs such as the Apple Newton, the machines used analogue mobile phone technology to send and receive email and faxes. Using the special electromagnetic stylus, users could mark up faxes on the screen before forwarding them to someone else or even attach voice annotation to sketches or documents before faxing or emailing them. Fax capability was still hugely important at the time, as most people had fax machines but few had access to email. So, with fax and telephone capabilities, the EO Personal Communicators were seen as having a distinct advantage over other proposed products. The company's brochure sold the product with typical hyperbole:

It's like everything else. You can fax with it. You can phone with it. Send an electronic letter, or receive one. Exchange information with distant computers. Communicate over the airwaves, without wires. You can do these things in ways and places that were never possible with the traditional tools—fax machines, phones, computers. And instead of many devices, you can use only one. We call it the EO Personal Communicator. Others have called it the future of communications. It may just be the biggest breakthrough since the telephone.[84]

The first model to market, the EO 440, was launched with a huge AT&T-funded fanfare at COMDEX in Las Vegas in November 1992. AT&T announced its vision of machines that would change the way people would work forever. Visitors could try out the machines, which became the talking point of the exposition.

As far as the audience was concerned, the EO 440 was a market-ready product, but in reality, it was still some way from finally shipping. The final code for the operating system was still causing problems for the software team at GO, and the relationship between GO and EO (never completely happy) deteriorated to the point where GO threatened to withhold the final version of the operating system from EO.[85] The revolutionary product finally made it to market in 1993,[86] but its bigger, better brother, the EO 880, was restricted to one hundred sales demo machines and never saw full production.[87]

AT&T, being historically a key player in the telecommunications industry, was particularly interested in the telephone functionality of these personal communicators and in the future predicted role of voice control in computers in general. AT&T, by now the majority shareholder, was driving EO to explore these aspects for future products. So while the engineering team was struggling to solve problems with the existing range, EO's management brought in design consultants to consider the next generation of EO products. One team asked to produce concept designs was from IDEO, including Naoto Fukasawa and led by Paul Bradley:

> We were asked to meet the CEO of EO several times. He was definitely someone who saw himself as a visionary. He asked us to create this product which was his personal vision for what the future might be—the idea was a kind of 'your own personal valet'.

Blue sky 'personal valet' concept for EO by Naoto Fukusawa, 1993.

Model of the AT&T EO Magni Personal Communicator, 1993–1994.

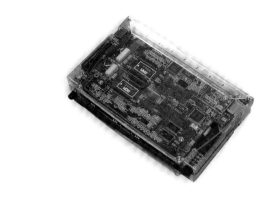

The team built a conceptual model of one of Fukasawa's designs:

> a combination of a pad with what we called a little valet communicator that was essentially your 'brain' and your phone. It was never intended to be real—it was a design test—and it kind of speaks of that time.[88]

Having passed the test, Bradley's team began work on a more serious product, the EO Magni. Its design semantically suggested voice control functionality by drawing attention to a 'half-moon' speaker, which Fukasawa referred to 'a kind of amphitheatre or acoustic catcher—like you were putting your hand around your ear, and I think that's what EO really liked about it—it really communicated that they were moving from a pen based device to a voice based one. In making the microphone more significant, the speaker really became the visual cue for that capability.'[89]

Another key feature was the large antennae—a requirement at the time for any wireless device. The product was quite bulky, as the casing contained a removable pen, a screen, motherboard and a large battery pack. The body of the unit was not flat but tapered to rise off the surface of a desk and had a lot of sculpting of the form in order 'to hide the thickness of the body'.[90]

The EO Magni got no farther than a model, although it was by no means a fantasy. It was based on 'pretty firmed up electronics. The intention wasn't to do a study—the intention was to go to market with this product. It was meant to be very real.'[91] However, other consultancies were also bidding for the next product with voice recognition. The concept finally chosen for further development was code-named 'Loki'. Ex-GO and EO product manager Henry Madden recalls, 'As cell phones got smaller we recognized that small is better and so we went ahead and made a smaller LCD display and "broke" the unit in half, folding like a book. It used the same technology as the EO 440 with less features but the market wasn't ready for such a high-tech product with hand recognition—it would have been very expensive for the average user.'[92]

In the end, the decision about which product to develop became a moot point. AT&T, disappointed in the non-appearance of the EO hardware and the GO software, considered switching to Apple's Newton software and even considered buying Apple outright. In order to keep trading, EO was merged back into a single company with GO, but the AT&T division responsible for computing products recommended closing EO down completely, which it eventually did in July 1994.

Clear acrylic prototype of Project 'Loki', 1994.

DESCRIPTION **A combined handheld PC and smartphone with colour LCD touch screen. Dimensions: 165 by 84 by 30 mm (6.5 by 3.33 by 1.25 in.).**

The DualCor cPC was a unique handheld computer product planned by a start-up technology company, whose founders' previous experience included setting up the first online pizza delivery service.

The cPC was a fully fledged tablet computer combined with a full smartphone and used two completely separate operating systems on two separate microprocessors, which provided a number of benefits over existing ultra-small computers. A small batch of working prototypes were produced and used to widely promote the product, and it was shortlisted for the Best of Show Award at the Consumer Electronics Show in Las Vegas. With several orders for the product, DualCor commissioned a redesign of the product to enable it to go into full production but failed to secure the necessary funding from investors, who couldn't see the difference between the cPC and existing products or see the potential market for it.

Publicity image of the DualCor cPC, 2006.

DualCor Technologies was founded in 2001 by computer engineer Bryan Cupps and businessman Tim Glass, who together had previously founded the first online pizza delivery service, Cyberslice Inc.[93] in Seattle in 1996. Inspired by the film *The Net* starring Sandra Bullock,[94] Cyberslice used the latest Sun Microsystems servers and software from Steve Jobs's company NeXT to translate orders into digital voice messages which were then relayed by telephone to participating restaurants.[95] Following Cyberslice, the pair set up an Internet advertising company called iWare in 1998 before finally starting DualCor Technologies in Scotts Valley, California.

Initially, the company was called GCV1, which meant nothing but could be guaranteed available to enable the company to be set up straight away. Tim Glass says, 'We felt we should wait to name the company until we had a better sense of who we were and how we wanted to be perceived.'[96] The early days of the company were taken up with engineering development work by

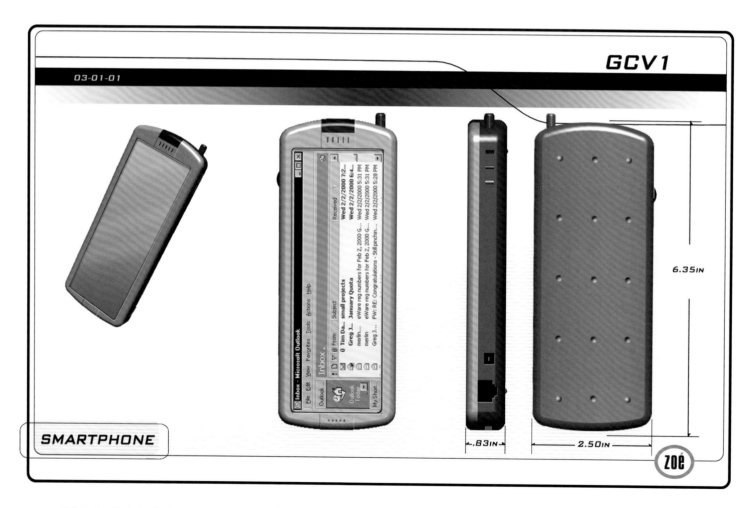

Kelly Kodama's design for the Chameleon, 2001.

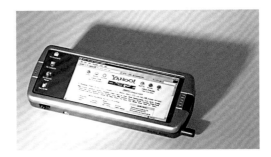

Appearance model of the Chameleon, 2001.

Bryan Cupps, who filed a patent[97] for a revolutionary 'dual core' system that formed the basis of their first product and led to the company's final name.

The project was internally called 'Chameleon' because the product switched between operating as a handheld computer and as a smartphone depending on the need, 'kind of like a chameleon lizard changes color to adapt to its environment'.[98] The concept was developed and prototyped as a proof of concept in 2001, with industrial design work by Kelly Kodama of Zoe Design Associates.[99] The original version bore a strong resemblance to a large mobile phone, but as the engineering became more resolved, the proportions of the product changed, and Kodama later produced a second design that was more square in format.

A third version based on these proportions, called 'Core', was developed early in 2005 with design work by Chris Loew. His design was based on the principle of laminated layers inspired by the Master Lock padlock. The unique construction was a stack of 2 mm material layers, with each layer's function paired with an appropriate material. The top dark layer reflected the screen and had perforations for speaker and microphone; the central elastomeric layers contained the connectors, controls and a post for a wrist strap; and the bottom layer acted as an aluminium heat sink. The layers were compressed together by a screwed post at each corner. A special double-tipped stylus could act as a stand for the device when placed into a slot in the base.[100]

As the product evolved over the next six months, various alterations were made to Loew's design in-house, in conjunction with a local consultancy, Function Engineering. 'The white antennae bump [sticking out of the left side] grew and shrank over time, the perforated metal front became a localized speaker grille, the control buttons moved to the front ... all typical project stuff.'[101]

The final device was named the cPC, with the small 'c' standing for 'connected'.[102] It was first announced through the computer technology news Web site *CNET* in December 2005, in an article with the tagline, 'It's a cellphone. It's a computer. It's the two invaluable companions of the modern executive in one.'[103] Though the cPC was originally aimed at a general consumer market, the CEO of DualCor, Steve Hanley, recalled telling Cupps, 'What you have here is genius, but it's aimed at the wrong people. This is for the global knowledge worker.' Consequently, Hanley told the *CNET* reporter, 'It will be aimed at sales representatives and executives who travel extensively.' The article pointed out that there had been previous attempts to introduce fully fledged handheld Windows computers (without mobile phone capabilities), but these had not sold well because they all suffered from short battery life and limited performance.

The DualCor platform solved the performance and battery life problems by running two completely different operating systems, Windows XP Tablet edition and Windows Mobile 5.0 Pocket PC phone edition, on two separate processors within the same device. Because Windows Mobile used far less power than Windows XP, the product could be used in full PC mode with a battery life of three hours, similar to a standard laptop. By switching to 'smartphone' mode and running 'lite' versions of applications, however, the battery life could be extended to between eight and twelve hours.

The *CNET* article sparked a huge amount of interest, and details were excitedly reposted on numerous technology blogs, at first accompanied by erroneous images of the wrong product. Many sites referenced DualCor's own immoderate press release, with one, *The Gadgeteer*, quoting,

> The DualCor cPC is the only all-in-one, wireless 'handtop' (handheld-desktop) computer that combines the power of a desktop PC, the instant-on convenience of a PDA and the always-connected functionality of a cell phone ... The DualCor cPC offers robust features and extreme flexibility for the mobile knowledge worker in supporting the traditional office environment as well as working at home, working from a client location, and working as a perennial road warrior.

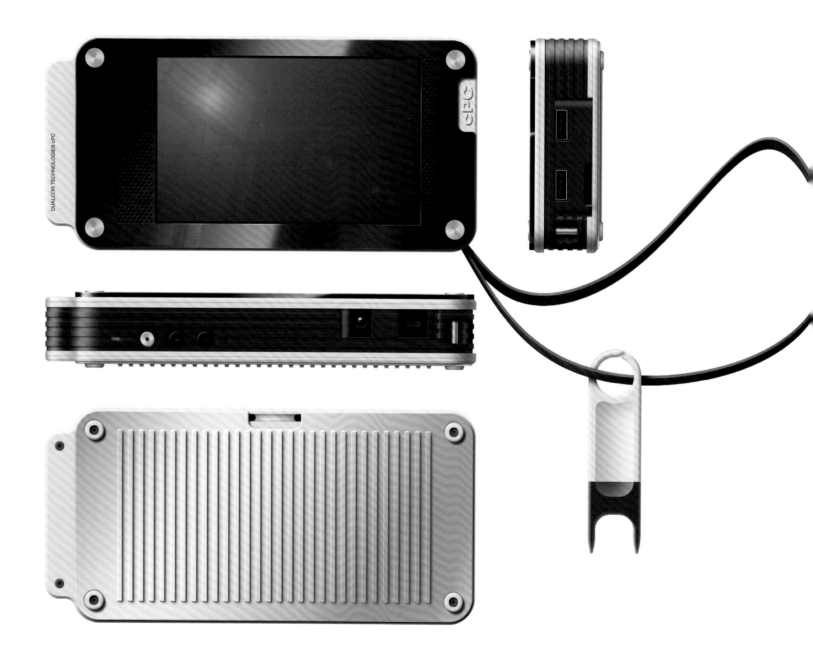

Chris Loew's redesign of the cPC, 2005.

It seemed to be too good to be true and raised questions about the reality of the product:

> Is the cPC a real product? If it is, then it might be the ultimate solution for business users, power users, travelers and true geeks. I for one would love to believe that it is not vaporware, but I am not encouraged by these photos, which look more like artist's renderings shown on the site.[104]

Another online technology magazine, *Gizmag*, ran a very positive article, saying,

> The DualCor cPC is a pretty special device and one which demands the attention of professionals who get out of the office a lot but don't want to forego the power of a desktop when they're travelling. This unique device represents a new category of hardware for global enterprise computing and promises to end the compromise between mobility, functionality and productivity for the mobile business professional.

The article went on to quote DualCor CEO Hanley as saying,

> 'The market response to the cPC has been overwhelmingly positive. Customers, partners and analysts alike have aligned in their collective praise and shared recognition of the cPC as a competitive differentiator in the world of mobile enterprise computing.'[105]

At the end of 2005, DualCor had a small number of 'engineering samples' made to show to investors and selected reviewers. One of these samples was demonstrated at the Consumer

The 'engineering sample' of the DualCor cPC with wireless modem, 2006.

Proposed redesigns of cPC by Peter H. Muller, 2006. (See also facing page.)

Electronics Show (CES) in Las Vegas, 5–8 January 2006, where it was shortlisted for a Best of Show Award,[106] and a video of the demonstration was posted online.[107] It had a 5-in. full-colour touch-screen display, 1 Gb of RAM, 1 Gb of flash memory and a 40 Gb hard drive, shared between the two platforms. A short stylus was clipped into a slot on the back of the device to use in tablet mode, or the user could choose to use the three buttons on the top, either side of the screen, as a joystick and two mouse buttons. A microphone, loudspeaker and headphone socket were all built in. The cPC had a slot for a wireless modem, three USB ports to allow it to be connected to a range of peripherals such as external keyboards, printers and scanners, and a VGA port to connect it to an external monitor, meaning the device could be docked and used as if it were a full-sized desktop PC. At this time, the device was expected to be priced at $1,500 and be available from March 2006.

Realizing that the design needed to be reduced in size and weight for production, DualCor commissioned Interform Ltd, an industrial design consultancy in Woodside, California, to redesign the sample into a version that could be fully mass-produced in volume. The founder of Interform, Peter H. Muller, worked on the project for around six months beginning in early 2006 and remembers that DualCor was quite well funded to begin with but needed further funding to bring the redesigned version into production. A problem arose when the venture capitalists

cPC Concept A
DualCor - 6.26.06

they approached were not overly impressed with the cPC, as it looked similar to other products that were already out there. 'They asked "What's the difference?" and the answer was "Yes, there's a difference, because it's a dual core system, not single core like the competition, and it's much more of a full-fledged PC". So they asked "Do people need a full PC?" and well obviously, smartphones are an answer'.[108]

Despite DualCor having agreements from 'several large companies and consulting firms'[109] to purchase units, it was unable to convince investors that there was a real need for the product and was cut off from funding it desperately needed. Muller thought it was a shame: 'I thought it was a very clever product. It had Bluetooth and an air card (a wireless modem). They were talking about a future version having a camera. No one had anything like this. I thought it was great, I would have loved it.'

Unfortunately, replacement funding was not found, and before long, smartphone technology had come to the point that there seemed little point in pursuing the matter. Tim Glass sees the circumstances around the failure of the cPC to reach market as an increasing problem for small companies. 'You just can't do projects like the cPC now. The small guys just get pushed out by the big corporations.'[110]

Mobile Computers

The desire to create portable computers goes back as far as the earliest real attempts to create personal computers—that is, stand-alone computers intended for use by one individual as opposed to a computer system shared between many users. Even before the technology became small enough and cheap enough to provide a portable machine with its own computing power, portability was a key issue, and many computer companies produced mains-powered portable terminals with built-in printers rather than display screens that could be moved from place to place and hooked up to a remote mainframe computer over the telephone network. In the early 1970s, these provided sales executives with the ability to send in details of customer requirements and receive back quotations for supply that could be printed out and immediately given to the customer. Slowly, small amounts of memory were added to these units, allowing a few pages worth of text to be stored, but it took a number of years before these remote terminals gained anywhere near enough computing power of their own to be truly called portable computers.

In those early days, the intended target market for portable computers was clearly the travelling executive. This was for two reasons: firstly, travelling businessmen (and they almost always were men) were really the only user group that at that time could justify the requirement to access a computer while away from the office; and secondly, because of the product development costs of cutting-edge technology and the initially low production figures, the first portable computers were always going to be highly expensive, and corporate executives were one of the few groups of users that could justify such an expensive outlay. These facts had been noted well before serious attempts were made to develop such a machine: an early prediction about portable computers by Honeywell was drawn up in 1966 as part of the research commissioned for Stanley Kubrick's *2001: A Space Odyssey*. They correctly concluded that in the not too distant future, portable computers would take the form of an executive attaché case.[1]

Although the aim from the start was to create a portable computing product with a suitably executive image, the first attempts to produce such a machine that appeared on the market looked less like a briefcase and more like a sewing machine that could barely be carried aboard the helicopters and private aeroplanes such people were portrayed as using. There is good reason why these machines were labelled 'luggables', as in reality they were only just transportable rather than truly portable, with even their creators saying that 80 per cent of them never left the office.[2] The brochures for these computers stressed their portability and functionality, but the products themselves clearly failed to portray the sophisticated, refined image of the upwardly mobile executive that was obviously required. Even when battery technology became good enough to provide a suitable power source, battery-driven portable computers tended to be units with comparatively limited functionality, a tiny amount of memory and small LCD screens that wouldn't drain the battery too much but only displayed around three lines of text. Though considerably smaller than a 'luggable' computer, they were still the size of a large writing pad, took some effort to carry around and lacked a certain cachet.

The first people to overcome this image problem did so by creating a radical product that took the form of the executive briefcase as its starting point, and in doing so created the product that became for many years the industry standard.[3] The GRiD Compass Computer, launched in 1982, was the first real laptop computer, as the term is understood today. It was the brainchild of John Ellenby, an ex-Xerox computer scientist who had a talent for seeing where emerging technologies were leading and the product possibilities that would transpire. Together with the industrial designer Bill Moggridge, Ellenby created a product that was so far ahead of the competition that it defined the form of the portable computer for all of its competitors. Unfortunately, it was so far ahead of the field and deliberately engineered to such a high standard that its cost was phenomenally high and it priced itself out of the market. Nevertheless, the form of the GRiD Compass lived on. As the cost of technology reduced over the coming years, portable computers reduced further in price and the image of the laptop expanded way beyond its original executive associations to become the de facto computer for a massive group of users. With a version eventually being produced by almost every computer manufacturer, the competition to create the most powerful, thinnest and lightest laptop still rages today, and a huge range of various different laptop machines act as everything from a replacement for a full desktop PC to the more stripped-down 'netbooks'.

So portable computers ceased to be the executive status symbols they were originally intended to be and became just another incarnation of the quotidian personal computer. However, there is a clear distinction to be made between portable computers and mobile computers that revolves around the latter's smaller dimensions, and it seems that only the smallest and lightest machines are now able to generate any level of interest or retain any element of status for their owner. While it is true early portable computers could be carried from one place to another, their size and weight were such that this was not easily achieved. Even today, a decent-sized laptop can weigh a considerable amount. In contrast, mobile computing devices are carried around by their owners as a matter of course, often without them being given a second thought. Where portable computers were housed in large cases of their own, truly mobile devices can fit easily inside a small bag or even a pocket.

Such ultra-portable or ultra-mobile computers as they are known today are handheld devices with almost the full functionality of much larger, desk-bound machines and can be seen as the result of a convergence process of a number of different products: the handheld computer, the electronic organizer or personal digital assistant (PDA), the interactive pager (which became the Blackberry email device) and the mobile phone. Depending on who was making them (i.e. a computer company or a mobile phone manufacturer), these products tended to stress one particular function, often to the detriment of others. This confusion over functionality and the lack of a clear product identity is possibly a root cause of many products that were developed and prototyped but failed to appear in what was for a time a complicated and extremely fluid marketplace for technological products. It was not until the launch of the Apple iPhone in 2007 that all these four aspects were sensibly combined into a single successful product and a coherent template for the next generation of computing devices was drawn.

Project **SINCLAIR PANDORA LAPTOP**   Designer **RICK DICKINSON**
Client **SINCLAIR RESEARCH LTD**   Date of design work **1985**

DESCRIPTION **A clamshell-design laptop, 300 by 170 by 35 mm (11.875 by 6.625 by 1.25 in.), with a flat-screen CRT display enlarged by fresnel lens and Microdrive tape memory.**

The opportunity to develop a small, self-contained mobile computer at a point in time when such devices were just starting to appear on the market was an exciting opportunity for Sinclair's design team that could have taken a very different path. The proposal for the Pandora Laptop brought together research and development work Sinclair had carried out in the course of developing some of its earlier, groundbreaking products. It was intended that the computer would store data on the tape-based 'Microdrive' technology first developed for the Sinclair Spectrum, and have a display consisting of the same 2-in. flat-screen CRT monitor the company developed for its flat-screen TV with a series of lenses and mirrors used to increase the display to a suitable size.

Unfortunately, using these technologies was something that the design team was not entirely happy about. They wanted to take advantage of newer, up-and-coming technologies such as floppy disk drives and LCD panels, but their hands were tied. Hence the name 'Pandora'—'it was a box full of all the things we didn't want'.[4]

Prototype of the Pandora Laptop, 1985.

Ever since Clive Sinclair founded his first company in the electronics industry, Sinclair Radionics, in 1961, he was fascinated with idea of miniaturization and producing simple, low-cost products. He developed a number of diminutive radios and hi-fi amplifiers, many of which sold in kit form, and as early as 1966 he developed the first miniature television, the 'Microvision', using a 2-in. CRT. Although this was exhibited as a working prototype and advertised for sale, manufacturing difficulties meant that the Microvision never entered production—a fate that befell a number of Sinclair products over the years.

One of Sinclair's great successes, though, was a range of pocket calculators first produced in 1972. The Sinclair Executive, for example, weighed only 2.5 oz (71 g) and measured 140 by 60

Sinclair ZX80, 1980.

by 10 mm (5.5 by 2.25 by 0.375 in.) and was advertised as 'the world's smallest and lightest pocket calculator'.[5] The clever, low-power consumption circuitry allowed the use of hearing-aid batteries and a consequently unprecedentedly slim injection-moulded casing by Clive Sinclair's brother Iain Sinclair. The result won the company a Design Council award. Five years later, taking some components from one of the calculators and combining them with a microprocessor, employee Ian Williamson created the Sinclair MK14—a simple microcomputer—and Sinclair entered the computer industry. From 1978, the computer was successfully sold in kit form through a new company, Science of Cambridge.[6]

In 1980, renamed as Sinclair Computers, the company launched the Sinclair ZX80. Based around the Zilog Z80 processor and having a unique vacuum-formed plastic casing and flat membrane keyboard, the ZX80 was sold by mail order as a self-assembly kit or a ready-assembled product for under £100. It was a huge success, with demand far outstripping supply. The company changed its name to Sinclair Research and quickly followed the ZX80 with a more powerful machine, the ZX81. This too was sold by mail order but was also the first Sinclair product to be sold in high-street shops (through newspaper retailer W. H. Smith). It was designed by Rick Dickinson, a recent industrial design graduate who had previously worked for the company during a student placement.[7] It had a high-quality injection-moulded case, with a similar membrane keyboard to the ZX80, and sold in huge numbers.

This was followed in 1982 by another Dickinson design, which became Britain's bestselling computer—the ZX Spectrum. This iconic machine, with its idiosyncratic rubber keyboard, was available with 16 or 48 kb of memory and from 1983 could be connected to a series of ZX Microdrives—tape-based storage cartridges containing a loop of 200 in. of magnetic videotape that could store up to 85 kb of data. Although not as fast as disk storage systems, they were considerably cheaper and significantly faster than the slow and unreliable compact cassette tapes used with many home computer systems. One problem was that the thin tape could snag, in which case 'all data are lost'.[8] This was the storage system chosen for the Pandora Laptop.

Alongside the development of home computers, Clive Sinclair had continued his obsession with miniature televisions. His unsuccessful 'Microvision' was followed in 1976 with a completely different design marketed under the same name but a different

Sinclair ZX81, 1981.

ZX Spectrum and Microdrives, 1983.

model number. The TV1A Microvision measured only 100 by 150 by 40 mm (4 by 6 by 1.5 in.), but it was massively expensive for the time. Nevertheless, demand was far higher than expected, but by the time production was ramped up to meet the backlog of orders, demand had slumped and the products were left on the warehouse shelf, causing a financial loss. An improved version launched in 1978 fared no better and caused Sinclair Radionics to be closed down in 1979 when financial support from the National Enterprise Board was withdrawn. Rights to make the Microvision were sold to Binatone.[9]

Despite these setbacks, Sinclair remained convinced that the right design of miniature television, at a low enough price, would be a hugely successful product. Research and development continued under Sinclair Research Ltd, with financial support from the National Research Development Corporation[10] and money from the sales of the successful computers invested in the development of a flat screen that would allow a suitably slim design. The flat-screen CRT used technology developed by the Nobel Prize–winning inventor of holography, Dennis Gabor, in the mid 1950s. Instead of the beam of electrons being fired from the back of the screen to the front, the electron gun was mounted at the edge of the tube, and the beam deflected 90 degrees onto the fluorescent back of the screen (known as a 'side-gun' CRT). Sinclair spent six years perfecting the tube technology and the special fresnel lens that was needed to correct the distorted image and make it large enough to be viewed comfortably. Eventually, the FTV1 Flat Screen TV (also known as the TV80) hit the market in 1984. The industrial design, again by Rick Dickinson, followed the black-box aesthetic of the ZX81 and was closely copied by other companies.[11]

Because of Clive Sinclair's aversion to LCD technology,[12] it was decided that the Pandora Laptop design would employ the side-gun CRT tube from the flat-screen TV, with an arrangement of lenses and mirrors as its display technology. This was despite various sources, among them *New Scientist* magazine, warning that the technology would soon be completely superseded by the LCD.[13] The design work went ahead anyway, with a number of different concepts being modelled, including one deliberately intended to look like a book. A working prototype was produced, but sadly, despite huge efforts to make the display work, it proved unsuitable. The lenses and mirrors made the image appear to hover somewhere between the front and back of the screen, and the narrow viewing angle meant that it could only be seen from

TV80 plus development models, 1984.

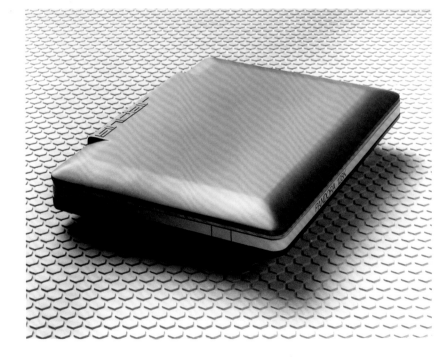

Models of different versions of the Pandora Laptop, 1985–1986.

Cambridge Computers Z88, 1988.

directly in front. If the viewer moved his or her head, the image appeared to move dramatically, causing a nauseous feeling similar to motion sickness.[14]

In April 1986, Amstrad bought out Sinclair's computer business. Alan Sugar took one look at the Pandora prototype, laughed and said, 'You'll have to do better than that, Clive!" With that, the Pandora project was dropped. Sinclair carried much of the technology on, eventually bringing it to market through a new company called Cambridge Computers in the form of the Z88 portable computer—with a notable lack of either the Microdrive or the flat-screen CRT.

Project **PHONEBOOK**
Client **VARIOUS**

THEREFORE PRODUCT DESIGN CONSULTANTS
Date of design work **1997–2007**

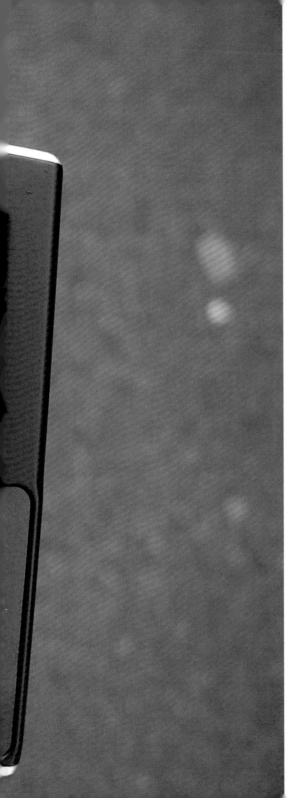

DESCRIPTION **A mobile phone and PDA which opens like a book revealing a full QWERTY keyboard and larger display to enable the easier use of additional organizer functions. Dimensions: 118 by 55 by 20 mm (4.625 by 2.125 by 0.75 in.) closed, 118 by 110 by 10 mm (4.625 by 4.375 by 0.375 in.) open.**

The Phonebook was an early attempt to combine the best features of an electronic organizer with a mobile phone and achieve a high level of typing usability within a small form factor. The idea was patented with the intention of licensing the design to a mobile phone manufacturer and was developed to the stage of a working prototype that could make phone calls. The design provided many advantages over that of existing mobile phones and attracted the attention of a number of different companies, but in spite of the praise and the promises, the design was not picked up by any of them. Eventually, the appearance of the smartphone rendered the concept obsolete, although it still has certain benefits.

Despite never having launched as a product, there has been misleading publicity about the Phonebook project. The online technology news site *The Register* ran a short story about the project but for some reason called it 'Datebook', and the prototype appeared in the book *Electric Dreams* by David Redhead,[15] which presented it as a fully manufactured, available product.

Working prototype of the Phonebook, 2003.

The Phonebook was a lengthy project originally instigated by Psion Computers and developed by Therefore Product Design Consultants over the course of nearly a decade. Psion Computers was one of designer Martin Riddiford's key clients while he was at Frazer Designers, and he had worked on the designs of their products since their early days. In 1993, Riddiford started Therefore with partner Graham Brett in a business partnership with David Potter, chairman of Psion Computers, and completed the redesign of its 1991 Psion Series 3 Electronic Organiser, the Series 3a.

The idea for Phonebook emerged from feedback from users of the Psion Series 3 Organiser, which revealed that the most important elements were considered to be the diary and contacts functions

The Psion Series 3a, 1993.

'Thor', an early version of the Phonebook concept under the Psion brand, 1999.

and the ability to enter typed notes. Certain groups of users, for example doctors, were also keen on the fact that it was outside of anyone else's system, so information entered into it could be guaranteed to be secure and private. That attribute would be compromised if the data resided on a computer rather than just on the handheld device.

It was while Psion was looking into the possibility of an online-connected PDA with mobile phone capability in the mid 1990s that the design team at Therefore thought about an alternative way of approaching the problem. They figured that there was great mileage in a product that had the diary and contacts functionality of a Psion Organiser as well as a mobile phone in one easy package, and could do it in such a way that it was easy to type information into it in a meeting for example.

At the time, getting all your contact details into the phone's memory was not an easy task—there was no syncing or commonality of formats, or anything like that. A number only keyboard [the letter-mapped phone keypad with three

Phonebook variants branded for Vodafone and T-Mobile, 2003.

letters per numeral] was a nightmare to use if there were lots of contacts to enter. We also knew from the original Psion Organiser that using a shift key to access alternate letters and numbers on the same keys was not useful. In fact, we called that 'Bastard Mode' it was so difficult. As we wanted to input addresses, we needed to be able to input letters and numbers at the same time.[16]

The idea was to somehow get full keyboard functionality into an object the size of a normal mobile phone without reducing the size of the keys to the point where they were unusable. Riddiford came to the conclusion that if the phone could open up like a book to reveal a landscape-format keyboard, there was no need for any buttons on the outside and so no need for a key lock. A user could open the unit up, type the first three letters of a contact's name, find the contact, press the call button and close the unit to use as a phone—all without any change of orientation.

Riddiford failed to convince others that a keypad on the outside was not needed, as it seemed to be a step too far, so a mock-up was produced of a phone with a standard twelve-button keypad on the front, a QWERTY keyboard on the inside and an animated user interface. A variety of design options were developed over the following year, including one that contained a shared scrolling roller to scroll through contact entries, which protruded through both sides of the front half of the case and so could be used both when the phone was closed or open. The code name for the project at this point was 'Thor', and the idea for the folding keyboard was patented internationally towards the end of 1997,[17] with Therefore due to get royalties from the sales of a finished Psion product.

One of the problems with the Phonebook project was the speed of change in the highly competitive mobile phone market, with new designs appearing many times a year. Given that at this point there was no real product, Therefore often found itself going back to square one and redesigning or refining the concept to take advantage of the latest developments (such as replacing the pull-out antenna with a newer built-in 'stub' antenna) to reflect the relentlessly shrinking size of mobile phones and keep the project up to date. A second version was developed for Psion code-named 'Pegasus', which was taken to the stage of full assembly drawings, showing the build proposal of having a battery in the front half of a magnesium casing and the circuit board in the rear. In the end, the roller idea was dropped as Nokia had a patent on the use of rollers, and Psion was nervous of litigation. Psion also required a finished product to ship with all the software included,

Phonebook prototype closed and being opened, 2003.

which would have taken too long and cost too much to develop. In any case, Psion was by then experiencing some problems as the market for technology products was going through a decline, and the company cancelled a number of projects, including a major pen-based smartphone joint project with Motorola called 'Odin'. By the year 2000, the company began to change direction, moving away from consumer electronics products, and had decided to pull out of the Phonebook project.

Convinced that there was a very real market for the Phonebook, Therefore decided to go it alone and invest in developing the product farther itself with a view to finding other mobile phone partners. A thorough development process was undertaken, with detailed over-centre hinge mechanisms developed to make the closing of the product feel right and a life-test rig set up to test the durability of the rubber keyboard during constant flexing. The keyboard layout was changed, replacing the row of numeral keys across the top of the QWERTY keyboard with a twelve-button number pad, as research showed that people found it easier to recall the 'shape' of a number keyed in this way. The black-and-white LCD was replaced with a colour version, and a camera was also added, placed on the front of the phone. This had the real advantage of being able to point in different directions, depending on if the phone was open or closed.

'Oyster' Phonebook for T-Mobile, 2004. The result of design by committee.

Therefore decided to license the design to a ventures company to raise funding for further development. A Far East manufacturer produced six samples of a prototype that had a mock-up of the user interface and could make real phone calls. A wide variety of different possible styling options were rendered, with different colour schemes, surface finishes and button layouts, branded for different companies; and working samples were produced with different aesthetics for European and Asian markets.

Over the next few years, Therefore made presentations of the Phonebook concept to a number of manufacturers, including Philips Consumer Communications, Ericsson, Siemens, Sendo, Motorola, Samsung and Blackberry. 'Every senior person we showed it to loved it—but you try and make it work in a company like that. Some people wanted to make it an Instant Messaging device, some wanted a great panoramic 3D thing—everything except a simple pocket address book in your phone.'[18]

The closest Therefore got to signing a deal for the Phonebook was with T-Mobile in 2004. During a breakfast meeting one morning, the head of mobile phones at Siemens was really keen to take the product, but Therefore had to turn him down because the company had a pencilled-in order for 150,000 units from T-Mobile, which was about to invest $1 million in developing a fully working prototype. The team from Therefore left the meeting with Siemens and went directly to a meeting with T-Mobile, where they were

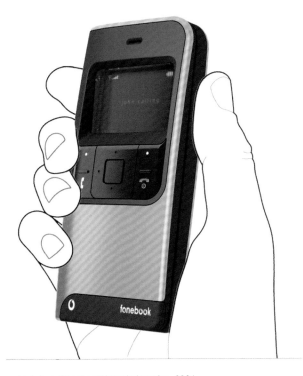

Vodafone 'Fonebook' branded version, 2004.

expecting to sign the final paperwork, only to find that as a company, T-Mobile had decided to work exclusively with Microsoft and the deal was no longer on the table. This in retrospect was probably for the best, as the input of different people from T-Mobile took its toll on the design, resulting in a version called 'Oyster' that Therefore was not at all happy with: 'Looking back on some of this, you think "Why the hell did we agree to do that?", but we were desperate to just go in the direction that people wanted it to go.'[19]

Following the T-Mobile debacle, detailed discussions were held with Vodafone, which funded some further development work on the design to prepare it for a large-scale presentation at Vodafone's annual convention of CEOs from around the world. Things were once more looking very positive when Vodafone suddenly announced the removal of a large number of senior staff: 'Literally two days before the show they ran a coach and horses through the management team, so all of these people were off or moved sideways into other jobs. So we sat there, with their investment in it as well, and they said "Well, there's no forum for this now."'

Such a knock-back was indicative of the project. In what proved to be a chequered history, Therefore kept the project alive for a long time without ever managing to clear the final hurdle. After the Vodafone experience, Therefore let the patent lapse in the mid 2000s. In the end, the Phonebook never did get made, and over the course of more than a decade, it cost Therefore Product Consultants a great deal of time, money and resources. It was not a complete waste though: 'The interesting thing about it is that we've earned from that project. Even though it never saw the light of day, our own patenting of it has led to other business—people have looked in on that patent and thought 'That's interesting, what those guys are doing' and here we are working for some of those people still after many years.'[20]

Project **PERSONAL INTERNET COMMUNICATOR (PIC)**
Client **SIEMENS AG, IC PRODUCTS DIVISION**
Designer **MARTIN RIDDIFORD/MARTIN WITTS: THEREFORE PRODUCT DESIGN CONSULTANTS**
Date of design work **1999**

DESCRIPTION **A clamshell-design combined PDA and GSM (global system for mobile communications) mobile phone with an adjustable angle, touch-sensitive LCD screen, full QWERTY keyboard and stylus. Dimensions: 200 by 115 by 30 mm (7.875 by 4.5 by 1.125 in.).**

According to its creators, the Siemens Personal Internet Communicator was 'the world's first Windows CE connected smartphone PDA'.[21] Developed in conjunction with a small, autonomous 'skunk works' team at Siemens in Munich, the fully working product was demonstrated at one of the largest consumer electronics trade fairs in 1999, and plans were drawn up to produce an initial product run of 10,000 units to be tested by Siemens staff.

The project failed because of software problems. The chosen operating system just couldn't support the required functionality that it should have been able to, and the product was quietly dropped.

Working prototype of the Siemens PIC, 1999.

The lead designer of the Siemens Personal Internet Communicator (PIC), Martin Riddiford of Therefore Product Design Consultants, had already had a great deal of experience working with electronic organisers, having designed Psion's first piece of hardware, the Psion Organiser, in 1984. Psion, founded in 1980 by David Potter, started as a software company with a close relationship to Sinclair Research, developing programs and games for the ZX81, the ZX Spectrum and the QL.[22] The Psion Organiser, with basic functions including a database, calculator and clock, was advertised as 'the world's first practical pocket computer', and with its launch, and that of its follow-up, the Psion II in 1986, Psion quickly became a leading player in the organizer and handheld computer markets.

The Psion Organisers were very successful products for a number of years but were made largely redundant with the appearance of the far more advanced clamshell-style Psion Series 3 in 1993, again designed by Riddiford. The advanced functions included full agenda and database packages, world clock and time zones, and word processing, spreadsheet and sketch drawing packages. It could also record sound and dial telephone numbers by generating the dial tones and being held to the mouthpiece of a telephone handset. The even more progressive Psion Series 5 appeared in 1997, boasting a touch-screen display and full keyboard with computer-style keys, which cleverly slid forwards when the unit was opened to counterbalance the screen.[23]

In fact, the design intention behind the Psion Series 5 was more aspirational than many realize. It was designed from the very beginning as an 'open platform' so that the LCD, keyboard and chassis could be used with different circuit boards, cases and other components. Riddiford recalls,

> When we did the Series 5, there was a whole new architecture created and there was a lot to do as it was Psion's third, ground up, completely new operating system and everything. The product was built in such a way, in 'stacks', so that it could be made 'mobile phone aware'.

The intention was that the next version of the Series 5, to be released in 1998, would have included a mobile phone modem, so it was built in such a way that it could have a thicker base to accommodate this. The plan was that after the version with the modem was launched, a later version would include full mobile phone capability, and after that, another, slightly thicker version would also include a full radio.

The Psion Organiser II, 1986.

But that never happened. Psion were then casting around looking for mobile phone partners because they didn't have the technical expertise in house to go and do the mobile phone bit.[24]

At exactly the same time, Psion, which had already split itself into a number of different operating arms, merged its Psion Software division with similar divisions from Nokia, Ericsson and Motorola to form Symbian Ltd. This was an attempt to exploit the convergence of PDAs and mobile phones that was starting to take place. The company developed Psion Software's operating system for the Series 5 into 'Symbian OS', looking to promote it as the preferred choice for all devices from mobile phones to multimedia communication devices.

Meanwhile, Psion's search for a mobile phone partner had resulted in a joint venture being formed with Siemens to develop a 'connective PDA', ergonomically combining a mobile phone and Psion's PDA. Therefore worked on around twenty or so different design concepts exploring the possibilities along these lines. The collaboration between Psion and Siemens didn't last though, partly because Psion's David Potter was only 'interested in the data

The Psion Series 5, 1997.

connection side of things, the email and web browsing, and not so interested in the voice communication side of things—phones needed different type approval. He just wanted an online device.'[25]

When Psion pulled out, the project was abandoned, but Therefore continued to work with the same group of Siemens people operating as a 'skunk works' team in Munich—a small, 'unofficial' group of half a dozen engineers working autonomously on product developments. Without the connection to Psion, there was no pressure to use Symbian as the operating system, and Siemens moved to an alternative. In Jim Fullalove's view, 'they wanted to be the first to show a data connective device using the new Windows CE platform, and they did it. It worked.' It was a far cry from a Psion product, though, and was a larger PDA product than Riddiford was used to working on:

> It was a funny thing, because at Psion, size was everything, absolutely everything, but Siemens weren't as concerned about size, as their computing background was in laptops. They were happy if it was used on a desktop rather than worrying if it fitted into a pocket. So it got a bit bloated as a product, as it had these great big batteries, a bigger keyboard and lots more electronics in it. But it did work—it was launched as a technology study, and they spent millions on it.[26]

The selected concept was based on an old design that Riddiford had produced while working on the mechanics for the Psion Series 5. It was a clamshell-type design that when opened, revealed an LCD touch-sensitive screen and a full QWERTY keyboard. Above the keyboard was housed a stylus to use on the screen, and the screen itself could be lifted up out of the casing to stand at various viewing angles if used on a desktop. A cylindrical section along the long edge of the casing opposite the hinge contained the large batteries used to give a prolonged use time between charges. The phone function was to have full 2G (second-generation) GSM capability using an internal microphone, and the unit was also to have a built-in radio.

Riddiford and Martin Witts worked on the form of the casing and produced plastic models with paper keyboards 'to understand the physicality of the product'. The design was fully developed as a 3D CAD model and rapid prototyped SLA models were produced to test the results before the mouldings were soft tooled. The internal components from Siemens were completed and the units assembled. The resulting working prototypes were not publicly announced, but they were 'quietly' demonstrated to selected

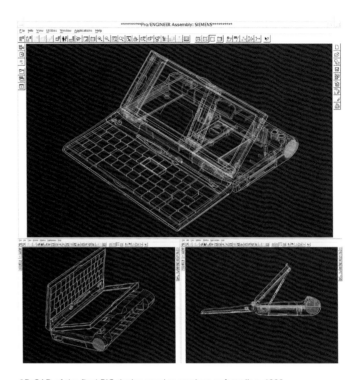

3D CAD of the final PIC design used to produce soft tooling, 1999.

audiences at the CeBIT Industry Trade Fair in Hanover, Germany, in 1999 as a 'technology platform'. Pleased with the response, Siemens planned to roll the product out internally, which was a bigger activity than might be imagined:

> Siemens employed about 330,000 people at the time, and they estimated that there were at least 50,000 technically savvy people who were internal candidates to use this device as a productivity tool, so they had a ready market for it. They intended to use the soft tooling to make a trial run of 10,000 to test the product in use.[27]

The final development of the product unfortunately didn't go to plan. Internal politics played its part. As Riddiford says, 'The

Studio photographs of the Siemens PIC, 1999.

trouble with all these projects is that they get tainted by the other people who get involved, moving the goalposts and trying to make their mark. It was as much a technology study as it was trying to get the design right and trying to get this Microsoft thing working, but it was compromised. They had written some of the software for it, but it wasn't properly supported by Windows CE.'

Windows CE was used in some other handheld computers in 1999, by IBM and Hewlett-Packard in particular, but these weren't trying to stretch the operating system with the addition of a mobile phone. In the end the Siemens product was let down by the software, which still at an early stage in its development was just not capable of doing all the things it should have been able to. 'Windows CE didn't work very well; it was just not up to scratch. It should have been able to email and surf the net as well as be used for hands-free phone calls, but it just wasn't usable.'[28]

The product got so close to being launched that a VHS tape press release was produced, ready to send to PR agencies, but the project was dropped before the tape was released. Nevertheless, even if it was never sold, 'it was the first mobile, portable device that was a PDA and a GSM phone.'[29]

Project **PSION HALO AND ACE** Designer **MARTIN RIDDIFORD, THEREFORE PRODUCT DESIGN CONSULTANTS**
Client **PSION PLC** Date of design work **1999**

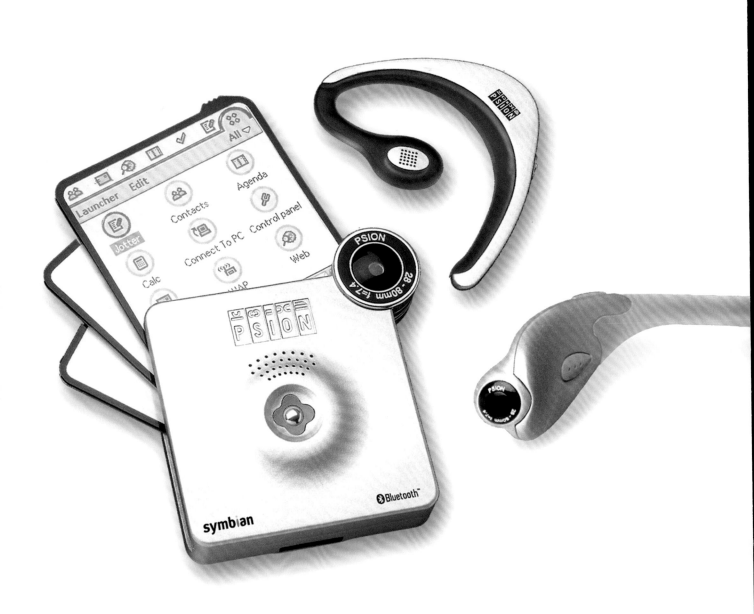

Psion Ace mobile phone/PDA and Halo headset concepts, 1999.

DESCRIPTION **Ace: a small, handheld mobile phone with built-in camera, Bluetooth headset and full PDA functionality, with a display consisting of three separate square LCD screens pivoting out of the body. Halo: a combined Bluetooth camera and micro projection system housed in an injection-moulded torc-shaped housing to be worn around the neck.**

The Psion Halo and Ace were a related pair of conceptual product proposals generated as part of the 'Psion Futures' project carried out by Therefore Product Design Consultants for Psion in 1999. The aim of the project was purely to generate publicity for Psion, which it did in spectacular fashion. Despite being only a series of block models supported by a promotional video, the Psion Ace and Halo designs were so realistic and convincing that they picked up a major award for product innovation from the consumer electronics industry.

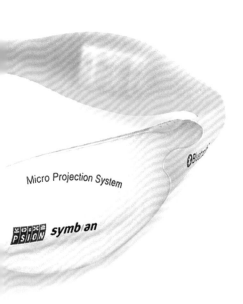

After a number of years producing innovative design solutions for Psion, Martin Riddiford's Therefore Product Design consultancy had developed a strong personal relationship with the company's founder, David Potter. Alongside working on the majority of Psion's successful mainstream product range of electronic organizers and handheld computers, Therefore produced a number of more futuristic concepts for Psion which did not go into production. In this vein, the Psion Futures project began as a purely promotional activity to promote the forward-thinking nature of the company. Around the time of Symbian being spun off as a separate company, there was a lot of associated publicity mentioning Psion, usually accompanied by photographs of David Potter. The problem was that every time the photographs appeared, they used stock archive images of Potter, more often than not with him holding a Psion Series 3 Organiser, which by that time was quite an old product.[30] Riddiford recalls,

Nokia and Ericsson and people like that were doing a whole bunch of future concepts and showing them at trade fairs such as CeBIT. At Psion, we were doing interesting stuff for real, but not showing any future concepts. So, we persuaded David Potter that we ought to be doing some future concepts

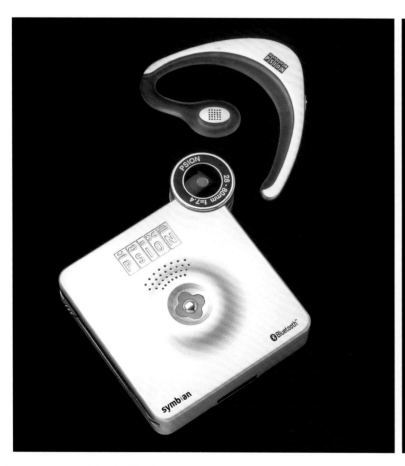
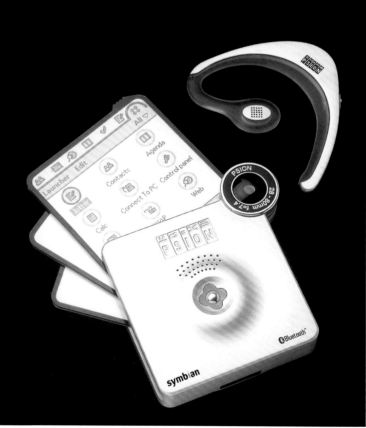

Concept model of the Psion Ace and earpiece, closed and partially opened.

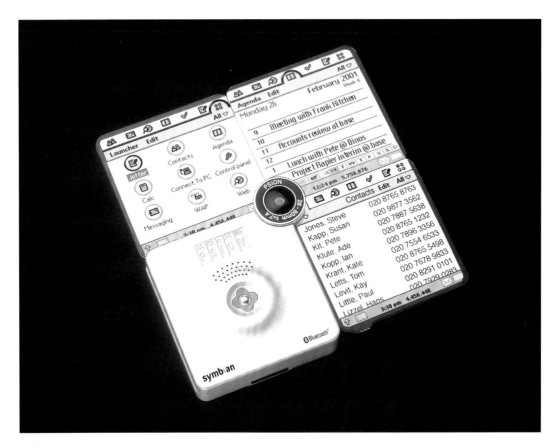

Concept model of the Psion Ace fully open.

to herald the arrival of this new 3G software stuff,[31] so that he could have pictures of himself with future stuff rather than past stuff, and he agreed.[32]

Therefore quickly developed concepts for a series of future products, took them to conclusion and produced hard models of them. They also made short videos about the products to explain their functions and how they would be used. 'They were very farsighted concepts, looking at how broadband would be delivered in the form of personal, wearable, new paradigm screen-based and non-screen devices.'[33]

One of them had an array of screens that splayed out like a fan so that you could run multiple apps on different screens. The other was a non-screen, projector-like thing that you wore around your neck and projected onto your hand, or a regular piece of paper. They were about ten years ahead of what was technically achievable at the time, but all potentially doable—things like mini projectors are available now.[34]

Riddiford explained that the concepts were 'heralding the notion of augmented reality'. The idea was that the user wore the 'Halo' device around the neck that consisted of a combined

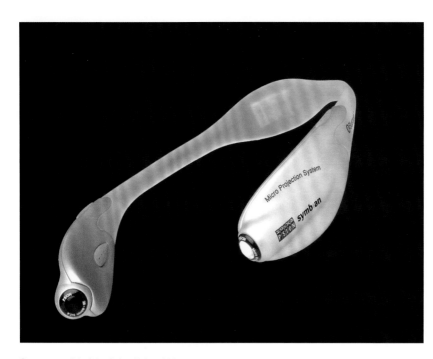

Concept model of the Psion Halo, 1999.

projector and camera. Depending on what was brought up in front of it, the camera would see it and automatically display the appropriate image. So, in a particular mode, if you brought your hand up in front of the projector, it would show you the address or information about the location, depending on where you were. If the projector was pointing at a small piece of folded paper, it would recognize the small scale and if it was in portrait or landscape format and display an appropriately sized image. Then if the piece of paper was unfolded, the image would increase in size accordingly until eventually 'the user ended up with a map or a tableaux or a newspaper or something, so there was a smart relationship between the media and the device'.[35]

With the three separate screens of the Ace fanned out, it was possible to have different, related apps running at the same time.

The user could, for example, have the agenda app showing the user's appointments for the day alongside their contacts list and an email program. Or the user could view Web pages on the Internet through a browser on one screen and have a calculator working on the second, while the third had a jotter app running so that notes could be written or recorded.

The concepts were distributed to press release agencies and presented at a number of international trade fairs:

> At one or two fairs that winter and spring, including CES and CeBIT, Psion, Symbian, Motorola were all punting these "roadmap concepts" and they got an enormous amount of publicity out of them, and they were simply conceptual models that we had packaged and made kind of "infomercials" for them as well to look at usability and connect them to a lifestyle, and it worked very well.[36]

Therefore was happily presenting the concepts at trade fairs and exhibitions when something unexpected happened. At the CES trade fair in Las Vegas in January 2000, the Psion Ace won an Innovations Award, as if it was a real product. The payback for this was even more publicity: 'Those concepts got absolutely loads of media coverage. For probably what was about three or four months worth of work, it was an extraordinary bit of PR for them.'[37]

Despite the success of the concepts, Therefore's relationship with Psion started to change. The publicity it generated was good, 'but I don't think in the way we imagined. It was a little blip for Psion—it said "Symbian has happened and here is Psion's vision of the future" sort of thing. We were assuming that there would be a body of work built up just like Nokia were doing but the press were cherry picking things a little and Psion went a bit flat after that.'[38]

According to technology journalist Andrew Orlowski, the reason Psion 'went a bit flat' was a direct consequence of the spin-out of Psion Software into Symbian Ltd. 'Psion found itself with no software engineers. In addition, the Psion Group's brains trust was broken up. Psion's ability to create integrated products, with close liaison between hardware and software experts, disappeared overnight.'[39]

For a while, new Psion products continued to appear, among them the Psion Series 7 sub-notebook computer in 2000. Development work on a new Bluetooth-enabled PDA code-named 'Conan' was underway when, the following year, Psion dropped out of the consumer electronics business altogether and began to concentrate on rugged mobile handheld computers for commercial and industrial markets—a market it had inherited through its purchase of the Canadian company Teklogix in 2000.

The legacy of the Psion Futures projects, though, is still out there. More than one Web site, including that of the BBC, carries images of the Psion Ace, with one stating it to be 'one of the five coolest British made phones ever'.[40]

An image of the Psion Halo projecting onto a user's hand.

DESCRIPTION **Aluminium-bodied notebook computer that could flexibly be used in laptop format or taken apart and rearranged into a desktop arrangement. Dimensions 350 by 255 by 50 mm (14 by 10 by 2 in.).**

The Dualworlds design was developed as an exercise simultaneously to solve the problems of poor ergonomics associated with laptop and notebook computer use and to remove the requirement for a desktop PC and laptop computer to be used in the same desk space. The solution cleverly combined the forms in such a way that in one configuration it was indistinguishable from other notebook computers, yet it could be altered to be ergonomically identical to a full desktop PC arrangement with separate mouse and keyboard.

The design was finished to a pre-production stage just prior to Compaq being taken over by HP, whose management had different ideas about the future markets for portable computers. A management decision was taken to cancel the project before production began.

Compaq Dualworlds notebook prototype designed by Morten Warren, 2001.

The Compaq Computer Corporation had a strong reputation for developing successful portable computers that went back to when the company was first founded in 1982. Compaq began as an idea by a small group of computer engineers from Texas Instruments to build a portable version of the hugely successful IBM PC. According to *USA Today*, Rod Canion, Jim Harris and Bill Murto met with venture capitalist Ben Rosen in a restaurant in Houston, Texas, where they sketched their idea out on a paper placemat. Rosen was impressed enough to back the venture and Compaq was born, with Rosen as chairman.[41]

The Compaq Portable was one of the first of the many IBM-compatible clones that eventually flooded the market and was the first successful portable IBM-compatible computer produced. It was designed and announced in 1982 and first sold at the start of 1983. It took the form of a suitcase-size box with a keyboard forming the lid—very similar to the 'luggable' Osborne 1 portable computer of 1981, itself closely based on Alan Kay's 1976 prototype design for the Xerox Notetaker. It was possible to produce a clone of the IBM PC because of IBM's decision to use so many off-the-shelf components and because Microsoft had managed to retain the rights to sell its operating system for the IBM, MS-DOS, to other computer manufacturers. The only part of the system Compaq had to create from scratch was the initial software used by the computer on start-up, known as the Basic Input Output System (BIOS). It is often stated that Compaq legally reverse-engineered a compatible version of the IBM PC's BIOS by using two teams of programmers—one to analyse IBM's code and turn it into note form and another team to turn those notes back into new code to achieve the same result. This, however, is not the case. Ex-Compaq employee Paul Dixon wrote that any computer programmers who had seen IBM's code were banished from working on Compaq's BIOS software. Instead, the team of programmers treated IBM's BIOS as a 'Black Box' of unknown content. They fed every conceivable command into the BIOS, recorded the output and then wrote code to replicate the same function, often far more efficiently than the original code. It took a year and reportedly cost Compaq $1 million to do this,

Alternative design for the Dualworlds notebook, 2001.

but the result allowed Compaq to legally produce an IBM clone and cleared the way for other manufacturers to follow the same route.[42]

The success of the Compaq Portable allowed Compaq to become the youngest publicly owned company to reach the Fortune 500 and generate $1 billion reserves.[43] The company quickly moved into standard desktop systems with the Compaq Deskpro range of computers, and over the next decade Compaq became one of the largest suppliers of PC systems in the world, with many large corporate clients, and eventually took IBM's place as market leader. During that expansion, Compaq acquired a number of companies and their products, including Tandem Computers and Digital Equipment Corporation, massively increasing the complexity of its product range.

With so many products under development, Compaq, like many other manufacturers, used external design consultants to bring fresh ideas to its existing research and development and design teams. One of these consultancies was Native Product Design in London, founded by Morten Warren. Native was given a fairly open brief to look for new product opportunities for Compaq.

Morten Warren recalls:

> The original brief stemmed from the fact that around 1999–2000, when notebook computers had become popular, Compaq's commercial customers, the huge banks and the oil and gas industries who bought thousands and thousands of pieces of equipment, discovered that issuing notebooks to their employees made them feel more valued, as it was almost like being given a company car, being one of the first to own these notebooks. And they discovered that they took up a lot of space, with the PC, the CRT, a docking station and keyboard, and people's desk spaces were getting pretty packed with stuff. So the project that was kicked off by the Houston division of Compaq, they asked us to start to investigate opportunities, go and meet some of their customers, interview customers and

Alternative designs for the Dualworlds notebook, 2001.

Compaq Dualworlds notebook, closed for carrying, 2001.

Compaq Dualworlds notebook, open in notebook format, 2001.

Side view of the Compaq Dualworlds notebook in desktop format, 2001

understand and gain insights into what was needed. It was very clear that the desk space had become very cluttered and that we needed to design something that had quite a small footprint. But the other thing that was quite clear was that the notebook was so well received, that changing the form factor of the notebook, i.e. maybe designing a desktop PC that somehow folded up and went into a bag, I think people viewed that as unattractive. We needed to maintain the form factor of the notebook and that's what Dualworlds tried to establish—to make a range of concepts but every single one of them enabled the user to maintain a notebook form factor, so there was no reason to reject the product. It could sit in a shop window next to a whole load of other notebooks and look completely at ease and look quite normal. It is only then, when the user interacts with it, that they find that the ergonomics and the whole function of it can change and it can hybrid into a desktop workstation. And that the footprint can be minimized, that's when people realized that the improved ergonomics ensured that employees' productivity was as good as it could be.[44]

Warren's team developed a clever form factor that would allow a laptop/notebook form factor to morph into a desktop arrangement. When first opened, the design was indistinguishable from other notebooks, but then the keyboard and the trackpad could be lifted out of the main body and placed on the desktop. The bottom half of the notebook could then be folded to raise the screen higher or in some designs raised vertically and held upright by 'pop-out' legs, and the screen itself could be angled out on a special hinged bracket.

The concept was developed to a highly resolved state in a number of materials and colourways to achieve a range of different appearances. Some of the designs had a very European, minimalist feel to them, but the preferred design had a very engineered, machined aesthetic, which appeared almost to have been milled from a solid block of aluminium. It clearly took design cues from photographic equipment flight cases, with infill panels in rigidized metal sheet to give a contrasting, 'more humanized' texture.

The Dualwords project continued throughout 2001 and into 2002, and during that time the project became more real, with working prototypes being built in preparation for production. Unfortunately, Compaq had by this time lost its top position in the market to Dell Computers, its share prices had fallen and thousands of staff had to be laid off.[45] Hewlett-Packard stepped in and bought the struggling company in 2002. Looking at the situation with fresh eyes, HP management took the decision to drop the Dualworlds project. Warren states:

I think that HP, when they took over Compaq, had a different viewpoint. Also the world had kind of changed—people realized they were actually very happy with notebooks, as they were starting to appear with bigger displays and by around 2003–4 we started seeing flat-panel displays starting to replace CRTs. However, I still maintain that had Compaq built Dualworlds I think it would have been a success. Everyone I've ever shown the concept to understands the immediate benefit. There are times when you've got a big project to do or a document that is very important and the ergonomics of working on a notebook, you just can't do it. Your body posture doesn't really allow you to be open and productive as when you have that distant operating view between you and the keyboard, the mouse and the screen by separating them out.[46]

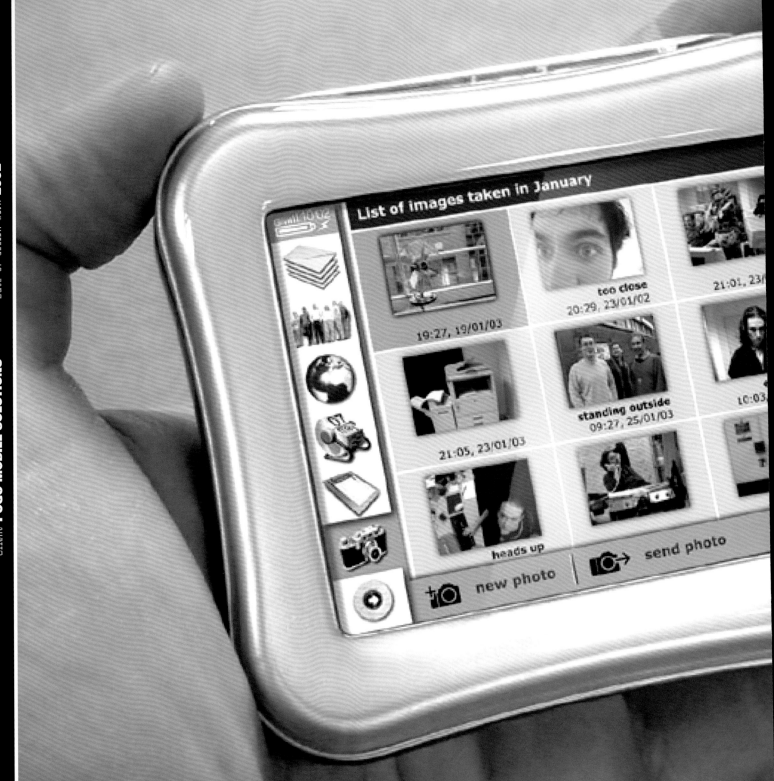

DESCRIPTION **Handheld mobile Web browser with combined phone, organizer, email device, camera and MP3 player. Plastic injection-moulded case with over-moulded rubber edges around a 3.5-in. touch-sensitive colour LCD screen. Dimensions: 100 by 85 by 16 mm (4 by 3.375 by 0.625 in.).**

In the early days of mobile telephone communication, the bandwidth necessary to wirelessly transmit and receive data meant that accessing Web pages on a mobile device was painfully slow. The technology to solve this problem, 3G phone networks, was announced as being just around the corner, but one company solved the problem a different way.

The Pogo's tagline was 'Why wait for 3G?' The device used existing 2G telephone network technology yet worked as fast as if not faster than the forthcoming 3G networks in which the mobile phone service suppliers had invested millions of pounds. No wonder they weren't keen to adopt it.

Prototype of the Pogo nVoy Communicator, 2003.

The original concept behind the Pogo emerged in 1999 from an internally run project at the London office of the US Web development company Razorfish, in response to problems with accessing the Internet on mobile devices. The limitations of second-generation (2G) mobile wireless technologies, which had been in place since 1992—and even its update, 2.5G (which improved transmission speeds up to 40 kilobits per second)—meant that handheld devices took an unacceptably long time to download standard Web pages, making for a very unsatisfactory user experience. It could be done; it just wasn't an effective way of accessing the Internet. Things were meant to improve with Wireless Application Protocol (WAP), a technical standard introduced at the end of the 1990s designed specifically for accessing the Internet over mobile devices, but as this used its own language, devices could only access specially created content rather than standard Internet pages.

Following some initial research, it became apparent that the majority of data in a Web page could be easily and heavily compressed. Razorfish developed the data compression technologies to do this, which allowed them to develop what was called a 'distributed browser' model for mobile Internet use—an early version of cloud computing. When a user wanted to access a Web page, instead of the device accessing it directly, it quietly sent a request for the Web page to the company's servers. The Web page was then located and compressed to one-fifth its original size, suitable for a small screen display, before being sent back to the device through the standard 2G GSM network at a speed that matched the dial-up modems used at the time in the majority of desktop PCs. The concept proved that Web browsing on a mobile device was a feasible proposal using existing technology—there was no need to wait for third-generation (3G) technology. This was obviously not welcomed by the majority of mobile phone service providers in the United Kingdom, who had all recently paid the government millions of pounds on obtaining licences in advance for operating the new, faster system. These providers were intent on developing new products to take advantage of 3G networks and were not interested in the idea.

However, in order to prove and promote the concept in a crowded marketplace, Razorfish needed a real product to showcase its revolutionary software-based solution. The company Pogo Technology Ltd was created, the data compression technology was patented and the product design consultancy Therefore was commissioned to create a 'funky, young product. Something different—a landmark crazy product'.[47]

Therefore's solution was a pillow-shaped handheld product that would combine Pogo's mobile Web browser with a mobile phone, an MP3 player and an email-messaging device. The radical form of the device accented the four corners, each of which had a particular function. One housed an on/off switch for the screen, one contained a thin pull-out stylus to write on the touch screen, one was the end of the built-in antenna and the last housed a socket for the mains charger. The unit measured 135 by 100 by 20 mm (5.375 by 4 by 0.75 in.) with a large 4-in. touch-sensitive screen surrounded by a rubberized 'bumper', which held a headphone socket and slots for a SIM card and memory expansion cards. Initially, this bumper was touch sensitive too, in order to extend the screen area. The screen display auto-rotated when using certain functions, and a small loudspeaker and a microphone to allow hands-free voice calling were built into the rear of the unit.

Using temporary soft tooling designed purely for prototyping, Therefore built a handful of working units in its workshops, and these were shown around to various possible partners before mobile phone retailer Carphone Warehouse agreed to undertake a trial. Impressed with the prototypes, Carphone signed an exclusive distribution deal with Pogo Technology at the end of October 2001[48] and placed an order for thousands of units, stating to the press that they would be available by the end of January 2002. Good product reviews by the industry piled the pressure on,[49] but without a manufacturing deal or even production tooling organized, the only way to meet this deadline was to use the temporary soft tooling to produce a full production run. A thousand or so units were hurriedly built in Therefore's studio workshop and delivered on schedule.[50]

The product concept was well received by the design industry, being described by the British Design Council as 'a compelling product solution'[51] and winning a Bronze iF 2002 product design award.[52] It also impressed the mobile phone industry, being selected as a finalist for the prestigious GSM Association Award for Best Wireless Handset/Terminal or Handheld Device and commended for technological advance in the Mobile News Awards 2002.[53] However, the Pogo didn't fare as well as was

(Facing page) Concept sketches of the Pogo by Marcus Hoggarth while at Therefore, 2000.

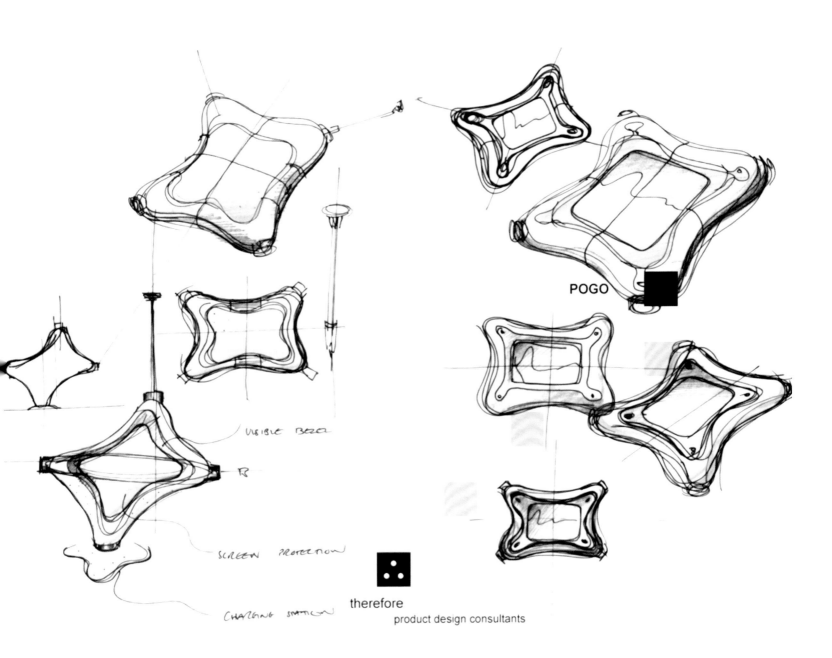

The original Pogo, 2001.

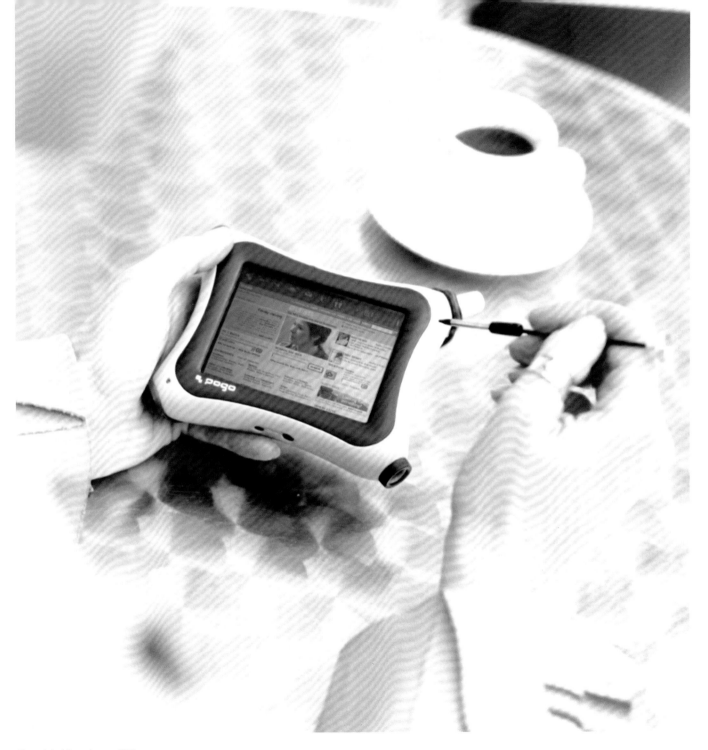
The original Pogo in use, 2001.

202   *Exploded view of the nVoy e100 prototype, 2003.*   *Working prototype of the nVoy e100, 2003.*

hoped in the marketplace. In many respects, it was a product slightly ahead of its time. The demand for Web browsers was not as high as it is today. It wouldn't sync to a PC like a PDA, which meant PC users couldn't update information locally but had to rely on their information being backed up on Pogo's servers. MP3 files were too large to download wirelessly through the GSM network, so music could only be played via memory cards. It came with some games preloaded, but there was a severe lack of third-party developed apps and games available for users. And finally, the removal of all buttons with just a screen interface was seen by critics as a just a step too far ('It doesn't have buttons—buttons are very important'[54]). Coupled with all of that, Carphone Warehouse would not subsidize the price of the device as they did with other phones, resulting in a steep price tag of £299 ($435), double the price Pogo Technology had hoped for. Unsurprisingly, sales were slow.

It was believed, though, that there was still some way to go with the product concept. So when Pogo Technology went into voluntary liquidation in December 2002, the technology and many of the staff were transferred to a new company, Pogo Mobile Solutions, headed up by a former Microsoft director, Ran Mokady. Mokady intended to license the technology to other manufacturers of wireless devices, as it would offer a much better experience than Microsoft's Pocket PC devices,[55] but needed a product that reflected the learning from the first Pogo and employed the very latest technological developments.

Marcus Hoggarth, the industrial designer responsible for the original design who had left Therefore Product Design to work at Pogo Technology and then Pogo Mobile Solutions, worked on the Pogo's planned successor, the nVoy Communicator. It was an attempt, he says, 'to turn a product that probably never should have been launched into a mass-market product'.[56] Considerably smaller and slimmer than the original Pogo and with a higher quality 3.5-in. screen, the nVoy was to be more square in shape than the Pogo's 'fat X' form, with scroll wheels in the two top corners to control different functions. It was designed to be a product that could be used in one hand, but it did have a short stylus clipped into its top edge to write onto the touch screen. Handwriting recognition software was vastly improved too, with the inclusion of a bought-in system called simpliWrite from Advanced Recognition Technologies.[57] The nVoy also had a camera built into the back of the product and an application dock on the screen to select functions. Web browsing was still to be a major aspect of the product, but a much higher emphasis was placed on email usage and texting, reflecting the network operators' view that this more 'mature' product would be of most interest to a business-oriented market. With this in mind, an alternative prototype, the nVoy e100, was developed that included a slide-out QWERTY keyboard, but this was not further developed.

A full-colour brochure of the final design aimed at mobile phone service providers and OEM suppliers was printed and distributed, but sadly, by the time the nVoy Communicator was ready for market, 3G phone technology had finally arrived in a blaze of publicity and new products and there was no interest from potential retailers. Before it could be launched, the nVoy had missed its window of opportunity and Pogo Mobile Solutions ceased trading at the end of 2003.

DESCRIPTION **A small proto-sub-notebook intended to be used in conjunction with a smartphone. The Foleo had a full-colour 10.2-in. widescreen, measured 268 by 169 by 24 mm (10.55 by 6.65 by 0.94 in.) and weighed only 1 kg (2.5 lbs).**

The Palm Foleo got so close to being shipped that numerous press releases and product reviews were published. The press release containing this image stated, 'The Palm Foleo mobile companion applications include email, full screen web browser, and viewers and editors for common office documents such as Word, Excel, PowerPoint and PDF files. The Foleo turns on and off instantly and features fast navigation, a compact and elegant design, and a battery that lasts up to 5 hours of use.'

Despite being a radical product, receiving widespread publicity extolling the product's virtues and being a computer that would have been the first of a new product type if actually sold, the Palm Foleo was subject to a very late decision to cancel production just prior to launch, leaving the initial run of computers to be scrapped.

Final working version of the Palm Foleo mobile companion, 2006.

Jeff Hawkins founded Palm Computing in 1992 after a period at GRiD Computers, where he had developed a number of new products including one of the first pen computers, the GRiDPad, in 1989. It was while observing users of the GRiDPad and canvassing their opinions that he noticed that the main things people seemed to like were the fact that it came on instantly rather than going through a lengthy start-up procedure like a laptop, and the fact that it had a very simple user interface. Users told Hawkins that they would like to store their own personal information on such a product, and these points led Hawkins to leave GRiD, form Palm Computing and work towards developing and producing the first successful personal digital assistant (PDA), the Palm Pilot, in 1996.

The history of Palm Computing is a convoluted one, involving a whole series of takeovers, sell-offs and mergers between different companies. In 1995, the company was bought by modem manufacturers US Robotics, who sold it two years later to the computer network hardware manufacturers 3Com. In 1998, Jeff Hawkins and fellow Palm founders Donna Dubinsky and Ed Colligan left and founded a rival PDA company, Handspring, where they eventually developed the Treo smartphone. In 2000, at the height of the Dot Com Bubble, 3Com listed Palm Inc. as a publicly traded company, losing much of its value when the bubble burst shortly afterwards. During 2001, the company formed a subsidiary called PalmSource to license its operating system, which became a separate company in 2002.[58] Palm then bought Handspring in 2003 and merged with it to form PalmOne. PalmOne bought the controlling share of the Palm trademark from PalmSource in 2005 and reverted to its former name, Palm Inc. After a decade of change, Palm was once again a single company able to develop both computer hardware and software.

One of the first products to be developed in this incarnation of Palm was another brainchild of Hawkins—the mobile companion—a small computer that would sync with and act as a larger screen and keyboard for a smartphone, specifically to accompany the company's increasingly successful line of Treo smartphones. As soon as it was turned on, the product would instantly display exactly the same email or office documents that were on the smartphone. Hawkins had the idea for this product when Palm joined with Handspring in 2003, but he was only spending part of his time with the business at that point and so hadn't had the opportunity to develop it.[59] All that was done at that point were an internal 'white paper' and presentations put together to support the project, which was code-named 'Hollywood'. Hollywood was to be 'the first connective experience between phone and computer—very ahead of its time'.[60] The Treo smartphone (which had a miniature QWERTY keyboard) was seen as a great product, but the view was held that 'people still need full screen and keyboards for composing lengthy emails, giving presentations or viewing attachments'. Although a laptop could provide this, it was by comparison slow and complex and required an Internet connection to email or browse the Web. The view was that the Hollywood device would be a success because wireless networks had matured and because there was a real consumer need for such a product that was not a laptop replacement.[61] Initial sketches of the Hollywood concept showed the imagined visual and tactile qualities of the product, made in aluminium and magnesium and having a rigid body akin to high-end aluminium luggage.

Once the design and development of the Hollywood project started in earnest, it took the best part of two years to complete. The software project was code-named MacGuffin and consisted of a customized user interface designed by Rob Haitani (the same person who worked on the original Palm Pilot interface), with Greg Shirai, built on a Linux-based operating system. The industrial design work of the Hollywood prototypes and the final Foleo were done by Palm's vice president of user experience, Peter Skillman. Some of the usability constraints had a strong impact on the final design: the product had to have a full-sized keyboard, which went from edge to edge of the device and so defined the width. As the intention was for the product to be carried in a bag, it had to have no sharp corners or details around the edges when closed; and so that it could be opened with a single finger easily, it had a magnetic catch rather than a latch. The fact that the device was designed to be held and carried was indicated by the rippled surface on the top of the product and the rubberized paint finish. To reduce the size of the device from front to back, a small trackpoint button and a roller for scrolling were used instead of the more usual trackpad. The roller was flanked on either side with left and right mice buttons and forward and back buttons.

The Palm Foleo was finally announced to the public at the end of May 2007 at 'D5: All Things Digital', the Wall Street Journal Executive Conference in Carlsbad, California, where Jeff Hawkins presented the product in person.[62] At the time of the launch, Hawkins said,

> Foleo is the most exciting product I have ever worked on. Smartphones will be the most prevalent personal computers

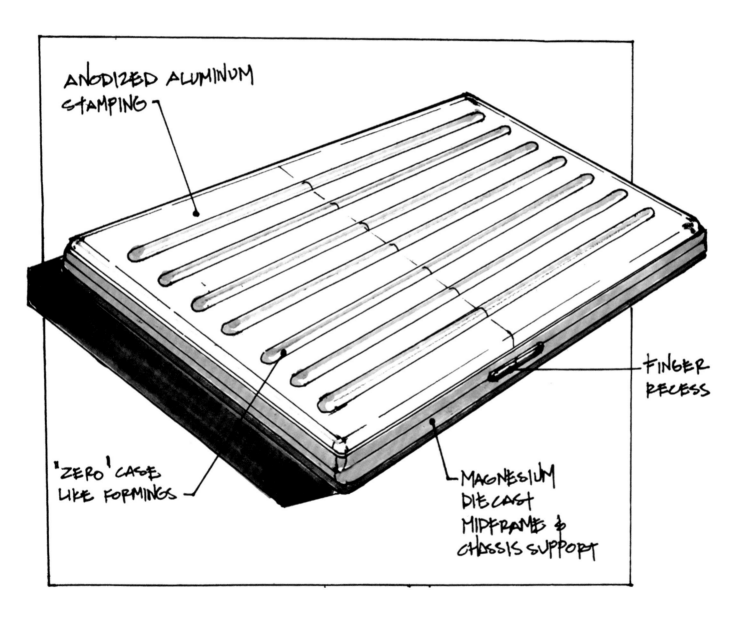

Initial concept sketch of 'Hollywood', 2003.

Final working version of the Palm Folco mobile companion open and closed, 2006.

on the planet, ultimately able to do everything that desktop computers can do. However, there are times when people need a large screen and full-size keyboard. As smartphones get smaller, this need increases. The Foleo completes the picture, creating a mobile-computing system that sets a new standard in simplicity.'[63]

Published reviews of the product were mixed. Comments were made about the price, seen as expensive at $499. Some journalists belied the advanced nature of the Foleo by comparing it with earlier sub-notebooks such as Compaq's Aero 8000 from 1999 or the Psion series 7 of 2000,[64] although these products did not have the synching ability or wireless connectivity that the Foleo had. Industry analysts were more cutting: following the Wall Street Journal Executive Conference, Gartner analyst Todd Kort said, 'I think it's probably the most disappointing product I've seen in several years. To think that anyone would carry something with a 10-inch display at 2.5 pounds as an adjunct to a phone just doesn't make any sense to me.'[65] Others were more appreciative of its novelty, labelling it the 'World's First Netbook'.[66] Hawkins admits that there was a lot of resistance to the product after its announcement. 'People said to me "I carry a laptop, so why would I need this? Why would I give up my laptop?" Really geeky guys didn't get it, but a lot of others—especially women—loved the idea because they could put it in their purse.'[67]

The Foleo was supposed to appear in stores in the summer of 2007, but it didn't. Rumours arose in the press that Palm was struggling, and the venture capital firm Elevation Partners was brought in to shore the company up. According to Jeff Hawkins, Elevation pulled the plug on the Foleo at the last possible moment. An official announcement of the decision was made by Ed Colligan, which stated that the company was cancelling Foleo 'in its current configuration' to focus instead on a new 'next generation platform' (WebOS) that would form the basis of its future products, including the Foleo II.[68] The decision to cancel the Foleo project at such a late stage was certainly not taken lightly. It was stated that it cost the company $10 million, as production lines had been set up, marketing materials had been made and a first production of 1,000 had been assembled, the vast majority of which were destroyed. Peter Skillman believes it was the right decision to make, given that there was 'not enough momentum in the existing ecosystem to support the product'. If, however, the Foleo had had the new Web OS, 'it might well have flown.'[69] Hawkins takes a more personal perspective, believing the reason it didn't ship was down to him taking his 'eye off the ball, and having lower than usual inputs to the project'.[70]

Palm did develop the new WebOS, launching it in 2009, but before it could create much impact, the company was acquired by HP in 2010. HP intended to produce a whole range of products based on the operating system, but after poor sales and strong competition in the form of Android OS products, HP finally closed the Palm hardware business down in 2011.[71]

The Agency of Ideas

The computing industry is often perceived, perhaps understandably, as a relatively recent phenomenon, almost, it could be said, one without a history. This may well be because our understanding of the word 'computer' is one that is constantly in flux, subtly altered with each technological development that becomes accepted as the current type form with which the public becomes familiar. Electromechanical computers, developed in the late 1930s, were more commonly referred to as 'calculators' rather than computers, and even at the time of the invention of the electronic computer during the Second World War, the term 'computer' was more usually used to refer to a human being than a machine.[1] If anything, computers then were referred to as 'Electronic Brains'. From the birth of the commercial electronic computer industry in the early 1950s and into the 1960s, the term 'computer' would for most people have conjured up an image of a room-sized behemoth: an array of large, grey steel cabinets containing thousands of glowing valves, programmed for a single calculation over the course of days or even weeks by a group of people connecting hundreds of patch cables into a particular configuration. The very same term, though, would have suggested a very different machine in, say, the mid 1980s, when almost every computer with which people came into contact was a personal desktop machine—an identical, IBM-compatible clone consisting of a beige box connected to a beige keyboard and beige mouse, with a large, beige CRT monitor as a display. For many people today, the term might suggest a laptop, a tablet computer such as an iPad or perhaps even a smartphone. At each stage, the history of the computer, for many people, is analogous to the history of the product the word means for them—i.e. going back only a few years at most. Before that, if anything, 'computers' were something else.

What is interesting to note is that where the computer in fact has a longer history than many people imagine, so too does vapourware, even if clearly it has not been referred to as such. Wherever there has been a computer of any description successfully marketed, there have been associated examples of machines that did not go into production at all. In fact, as might be expected given the complexities of the computer's development, vapourware predates actual computers.

THE CAUSES OF VAPOURWARE

Vapourware, it would seem, is a more complex issue than might be imagined. On the face of it, it has been understood by many as a simple case of companies merely announcing a new product in the full knowledge that it was not and might never be available for purchase. As a result, it has been considered as an inherently negative phenomenon—at best a case of overambition coupled with an amount of incompetence, at worst a deliberate, sometimes malevolent attempt to gain benefit from misleading others. From the evidence, though, it is clear that there are numerous factors that can result, more often than not inadvertently, in computers not making it to market. Furthermore, it appears that it is quite usual for more than one of these factors to be in play at one level or another at any one time.

Two of the most common causes of vapourware, financial problems and technical issues, are almost always in interplay with each other to some extent. Either investors withdraw support or executives decide to cut budgets because technical difficulties prove difficult to overcome; or companies invest so much time and resources into solving such problems that money finally runs out and receivers are brought in. In other cases, there is at least an amount of foresight involved when funding is refused in advanced because it appears that the amount of time, effort and resources to take a concept through to realization are just too great to make it worthwhile. In these cases, though, the dividing line between foresight and being risk-averse is a fine one. Looking in more detail at these different factors reveals some interesting subtleties and highlights the complex intertwining of the issues involved.

Charles Babbage spent the last fifty years of his life in an attempt to perfect his Difference Engine. The generous funding he had received from the government was cut off when after twenty years he had only managed to produce a demonstration model of a single section of the machine. For the remaining thirty years of his life, during which he attempted to improve the design of the Difference Engine and develop the even larger and more complicated Analytical Engine, he relied on his own financial inheritance. Babbage's financial difficulties in realizing his dreams of a difference engine occurred because of the interminable delays that were largely due to his own perfectionism. At every stage, he would rethink and redraft components as he saw improvements that could be made, marking his revisions on existing working drawings and 'often leaving the machines imperfectly specified or in transitional states of incompleteness'.[2]

A century later, a similarly lengthy development process caused problems for Sønnico in their attempts to realize the Hofgaard Machine. A promising patent on a theoretical principle took decades of detailed experimentation to turn into a working prototype—hampered all along the way by the perfectionism of its constructor unwilling to delegate any aspect of its operation. In the process, the project seriously overstretched its patron company, taking it to the brink of insolvency. With Hofgaard and Babbage, procrastination turned out to be the project's undoing.

More recently, the fledgling Phonebook project similarly started with willing financial support from Psion, which was withdrawn when the company experienced financial problems. The Phonebook was not a technically difficult project to resolve, but it nevertheless required significant software development. Therefore Product Design Consultants had to financially back their convictions themselves for much of the project's life, costing them around £120,000, which 'for a small consultancy is quite a significant bet'.[3]

The GO Computer is another example of a convoluted and troubled development process that entailed the spending of huge amounts of investors' funds in trying to produce a successful working prototype before finally giving up the ghost. Even the financial backing of some of the biggest names in the computer industry failed to result in a marketable product. Of course, funding is not only withdrawn or withheld before a product is fully developed. Jeff Hawkins's pet project, the Palm Foleo, went through a relatively smooth development process resulting in an initial production run of fully working products. It was only at this point when, fully prepared for a major product launch, financial pressures within Palm brought in new investors who were not confident in the market for the product in its current form. Exactly the same thing happened to DualCor Technologies with their fully working production model of the cPC that only required productionizing yet failed to convince investors of its potential.

A number of the case studies presented failed to reach the marketplace because of new technologies not being fully realized or stable enough for a consumer product. The IBM Aquarius project utilized a largely untested type of solid-state memory, 'bubble memory', which held a great deal of promise but, it was considered, could not have been perfected within a reasonable timescale or cost. Similarly, Sinclair Research's foray into wafer scale technology had its own problems that made it uneconomic for high-volume mass production, while its attempt to increase a flat-screen CRT to computer display size needed far more refinement before being usable. The already mentioned GO Computer was not only trying to solve the technical issues around electronic ink but also attempting to create a completely new form of gesture-driven operating system. A follow-up project, the EO

Magni, was intended to use voice control, which had not been adequately resolved into a fully working system. In contrast, the Siemens PIC used an existing, fully functioning operating system in Windows CE, but by adding phone capability, it was stretched to its breaking point.

Alternatively, some products did not make the market because, either before or by the time they were finally ready, the market had moved on and the window of opportunity had passed. Early examples include Nordsieck's Analyser and the Hofgaard Machine, both of which exploited technology that was already old before prototypes were built and both of which were badly affected by the introduction of newer technologies that rendered them obsolete. It is a problem that has become more and more of an issue as time has passed, as the pace of technological change has increased. The flat-screen CRT of Sinclair's Pandora stood no chance of reaching the market once LCDs took hold. Similarly, some of the benefits of Compaq's Dualworlds laptop, namely reduced desk space, were largely removed when large, bulky CRT monitors began to be replaced by flat-screen LCD versions. Finally, the introduction of improved mobile phone technology has hit a number of projects. The introduction of 3G connectivity killed the Pogo nVoy before it could be launched, and the wide acceptance of the smartphone after the year 2000 sounded the death knell for projects such as the Phonebook and the DualCor cPC.

Another key factor that is often a cause of vapourware is the internal politics involved in a project. The relationship between those promoting a project and those financing it, whether internal or external, is crucial. Internal politics in large corporations is particularly problematic, with IBM and Xerox being eminent cases in point. It seems particularly ironic that the very corporations that could afford the resources to attract the best people in the industry and support them for significant amounts of time in developing radical products that might have changed the future direction of the industry are the very ones that seem to have been most risk-averse when it came to supporting the projects into production. For years IBM's Executive Board refused to accept that it should even entertain the notion of a personal computer, and as a result IBM missed out on the chance to be the creator of a home computer industry that turned out to dwarf the business industry it had helped found. Xerox too was so focused on timeshared and large-scale business computers that it saw no advantage in trying to realize Alan Kay's dreams of a personal machine so simple it could be operated by a child. Many other projects have suffered through a lack of management support, differences of opinion between company founders and the executive boards they end up working for, disputes between internal departments wanting to go in different directions and push different pet projects, or poor relationships between in-house design departments and high-profile external design consultants.

In his 1999 book *The Invisible Computer*, Donald Norman put forward an argument that 'a successful product sits on the foundation of a solid business case with three supporting legs: technology, marketing and user experience'.[4] Using a diagram of a three-legged stool, his point was that if any one of these legs is not securely in place, the other two alone cannot support the product and it will fail. Vapourware, as has already been established, does not get exposed to the same acid tests of user opinion as products that make it to market, but we can nevertheless start to construct a comparable model of the supporting pieces which need to be in place for a design not to end up as vapourware. Here, though, those supporting structures are financial, technological and political. Financially, as many have found to their cost, a solid financial foundation is crucial. Money is required at every stage of a product's development, and many very worthwhile products have not made it to market purely because of cash flow problems. Technologically, a product might not necessarily have to provide the best ever user experience, but it does have to be stable enough or resolved enough to enable a reliably working product to be manufactured in quantity at a worthwhile cost. Of course, because of product development timescales, if a technology is too stable, too mature, it runs the very real risk of being made obsolete before it reaches the market. Procrastination can kill, and as in comedy, timing is all. Politically, a product has to have the full support of all involved. A conservative executive board that is risk-averse, a corporate culture of caution, investors who fail to see a product's potential or management teams that would rather stick with what they know can all too easily pull the plug on an otherwise promising project. Changes of management are also fraught with difficulty for products in development. As one company is taken over by another, it often brings a change of agenda, a new broom sweeping away the detritus of old projects, budgets cut completely purely because different regimes have different ideas.

It seems clear, then, that the pieces of the jigsaw that have to fit together to enable a real product to appear at all are every bit as important, complicated and involved as those that need to be in place for a product to succeed once it finally makes it onto the market.

WHAT IF?

It is a moot point, perhaps, to speculate on the impact a particular product would have had had it made it to market, but nevertheless, it seems fairly likely that some of the case studies discussed, had they gone into production at the time of their design, might have had a significant and lasting effect, possibly changing the face of computing itself. The potential impact of Babbage's mechanical computers becoming commonplace by the dawn of the nineteenth century has long been discussed, imagining a world where mechanical devices reached the complexity and capabilities of electronic devices today. The concept even formed the basis of the first steampunk novel, *The Difference Engine* by William Gibson and Bruce Sterling, which launched a whole subculture.

Science fiction aside, speculation on the potential effects of vapourware has its attractions. What if IBM executives had said yes to the Aquarius personal computer prototype? The radical solid-state technology involved might have been developed to a stage where it became the industry standard, in which case we might never have seen the 3.5-in. floppy disk and internal hard drives might never have been as ubiquitous as they are today. Added to which, we would have had personal computers a number of years earlier, IBM might have taken the lead in developing the home computer market instead of playing catch-up, and they might have still been the dominant force in computing today. Similarly, Xerox might have been a leading exponent of cutting-edge computer technology had they supported and put the Notetaker into production or pushed the development of the Dynabook instead of letting other companies take the benefit of much of their research.

As for pen computing, if that had proved to be the panacea it promised to be, the world of computing might look a very different place. The various designs suggested all pointed to an interaction experience with computers that still looks attractive today and might have led to the appearance of the gesture-based multi-touch screen as used on the iPad many years earlier than it finally appeared. It may also have directed attention away from the PDA and consequently the smartphone. Who could possibly say?

So, although we can speculate on what might have happened had a vapourware product made it to market, what is obviously far more important is what effect it did have despite the fact that it didn't make it.

THE IMPACT OF VAPOURWARE

One of the stated aims of intentional vapourware is to promote the company making the announcement, either purely to attract attention and keep the company in the news, to divert attention from competitors or to encourage customers to wait for a much delayed product. This is not always as negative an act as has been presented, as can be seen in the case of the Psion Halo and Ace project—a deliberate and overt publicity stunt which, like the Honeywell Kitchen Computer, was never intended to seriously mislead the public. Picked out as an award-winning innovative product, the project attracted far more attention than anyone had hoped and is still seen by some as a real product proposal that should have gone into production. Therefore's own Phonebook project was a serious entrepreneurial attempt to develop a new product, but it had the unintentional effect of promoting the industrial design skills of the company and bringing in work to the consultancy by exposing it to a new audience of potential clients.

Other pieces of vapourware have had significant impact within the companies involved. The IBM Atari PC, for example, was a deliberate move to persuade executive directors to give permission to rush the IBM PC project through to production. Without that provocation, IBM might have taken years longer to get into the personal computer market and missed the opportunity altogether. The IBM PC proved to be one of the most successful computer designs of all time and became the industry standard for many years. But this was largely due to the fact that, because of its rushed construction from existing parts, others could so easily copy it. Unfortunately, the only beneficiary of all the copies was the company that had written the operating system it used and retained the rights to license it to others—Microsoft. As it turned out, the success of the IBM PC proved to be the thin end of the wedge in the decline of IBM's fortunes as it lost its dominance in the industry.

Clearly, the fact that vapourware doesn't get to market does not mean that it is of no importance or has no impact. Although a particular product, for whatever reason, may not be put into full production, it is often the case that the research and development work involved in its creation remains completely valid and ends up in one form or another in a different product. Possibly not one as exciting or as potentially market changing as the original but, nevertheless, one that results in an actual product for sale. This was the case with the 'desktop' Saab D2, the components and architecture of which formed the basis of a range of much larger mainframe business computers. The cramming of technology into a portable case in IBM's SCAMP found an outlet in the very successful IBM 5100 computer and its derivatives and later informed the design of the IBM PC itself. Sinclair Research's work on wafer-scale technology finally appeared on the market in the form of spin-off company Anamartic servers, whereas much of the work done for the Pandora laptop was reused as the basis of the Cambridge Computers Z88. The Apple Figaro, too, saw the development work for the full product diverted into a smaller, far less ambitious product that eventually became the Apple Newton MessagePad, popularizing the new industry of the PDA.

It is often said 'While failure is an orphan, success has many, many parents'.[5] The point is that whereas people are more than happy to highlight their role in a success story, they are usually less forthcoming to admit their failings. Whether orphan or not, vapourware nevertheless has a history of pushing technological boundaries and as a result inspiring and motivating others—often, but not always, resulting in far more success. There are various examples of such impact that show this to be a consistent element, going back to the very roots of the computer and the work of Charles Babbage. From a simple engineering point of view, it is acknowledged that Babbage's demands of his master engineer John Clements 'led to advances in precision tool and parts manufacture in England'[6] and to Babbage becoming a respected authority and author on the subject of manufacturing,[7] but his influence went much farther.

Babbage's attempt to build a difference engine was widely known during his time, being disseminated through popular and academic articles and consequently inspiring many to follow in his footsteps. One such follower was the Swedish printer, publisher and inventor Pehr Georg Scheutz, who in 1834 read of Babbage's work in the *Edinburgh Review* and was driven to create a working difference engine of his own.[8] He designed the Scheutzian Calculation Engine in 1837,[9] and although he initially believed one such device would be all the world would need, he later hoped to sell the machines in quantity. With his son Edvard, he developed a working prototype by 1843 and two fully engineered versions in the 1850s, although his unreliable machine was of far lesser capability than that proposed by Babbage.[10] When the Scheutz engine was sold to the Dudley Observatory, another Swedish printer who wanted to use the machine to create interest tables, Martin Wiberg, was forced to develop his own version, which he did in 1860. Although, as for Scheutz, it was a financial disaster, Wiberg did receive 800

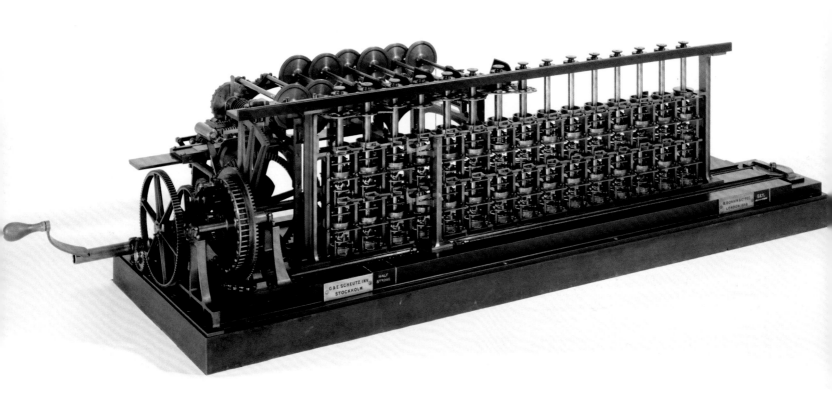
The Scheutzian Calculation Engine, 1853. Directly inspired by Babbage's well-publicized work.

kronor from the Swedish government, several medals and awards and the recognition of Napoleon III for his trouble.[11] These were not the only ones inspired by Babbage to produce a difference engine. In the USA, George Bernard Grant produced a working prototype of a large machine that was exhibited at the Centennial International Exhibition at Philadelphia in 1876, though he decided to take it no farther.

Long after his death in 1871, and throughout the twentieth century, the work of Charles Babbage continued to influence that of others—largely through the efforts of his son Henry Prevost Babbage, who travelled widely, promoting his father's achievements through public lectures and in 1889 publishing a full account of the development process involved in *Babbage's Calculating Engines* (a book which is still in print).[12]

The second case of inspiration, the Honeywell Kitchen Computer, had more of an effect than either its manufacturer or promoter imagined. Even though it was never intended to sell as a real computing product, it presented Honeywell as a cutting-edge technology company and enabled it to better compete in a rapidly changing marketplace. As a fantasy gift, it caught the attention and imagination of a curious public, raising awareness of the Neiman Marcus brand exactly as intended. But it also had some unintentional consequences. It caused other manufacturers to seriously consider the 'what if' scenario of a computer in the home and acted as a stimulus to a whole process of research and development looking at potential products for a new market. The vice president of Digital Equipment Corporation, Gordon Bell, urged the executive board to take this notion seriously in 1969, but those above him, including founder Ken Olsen, were famously opposed to such ideas. Bell went on to develop ever-smaller minicomputers for DEC and was eventually proved right when the home computer market exploded in 1977. DEC didn't respond until after IBM launched its IBM PC in 1981, at which point the company did produce a range of more powerful (and more expensive) microcomputers. Unfortunately, they were not IBM-compatible (and not even compatible with each other) and so were not successful in the marketplace.

Thirdly, Alan Kay's concepts for future computers produced during his time at Xerox PARC had perhaps more impact than most. The wide dissemination of his Dynabook concept to leading thinkers in the field through academic papers is widely recognized as having given direction to numerous research teams both within and outside Xerox. It was truly a vision of how much computing technology could achieve in the right form. The work done by different teams within Xerox, all sharing the same intellectual space as Kay, might have had a chequered history in terms of real products that were successful in the marketplace, but there is no doubt that, without them, much of what we accept as standard technology today would not exist.

On a more pragmatic level, Kay's 1975 prototype portable computer, the Notetaker, may have had a number of issues of cost and weight that made it an unattractive proposition for Xerox to put into production, but it was a clear enough indication of the near future of computing to be taken up by others. The first to do so was Adam Osborne, who, with Lee Felsenstein and Jack Melchor, took inspiration from Kay's design, founded Osborne Computer Corporation and in 1981 very successfully marketed what many have labelled 'the first commercially-successful portable personal computer'.[13] Despite its obvious disadvantages as far as being portable went, in the face of no competition at a reasonable price, the Osborne 1 satisfied a latent demand for portable computing few had realized was so strong. The success of the Osborne 1 created a flood of followers who, over the next few years, released variations on the same basic theme, each with his own improvements. These included the Kaypro II in 1982, which had a much larger display screen and could store much more data; the Compaq Portable in 1983, which had the huge advantage of being IBM-compatible; and the Commodore SX-64 in 1984, which although it had a smaller display was full colour. Among them, these products defined the form of portable computing until such time that technological advances enabled the more expensive but eminently more suitable form of mobile computer—the laptop—to make them not only technologically obsolete but also stylistically anachronistic.

The fourth exemplar case of a piece of vapourware inspiring others is the prototype pen-based GO Computer. One of the best-documented cases of vapourware known, GO promoted a whole new world of computing products in which people would write onto computers as easily as using a notepad, and for a number of years, the intention was accepted as the future for computers with complete certainty. It was commonplace throughout the early 1990s to see headlines on computer magazine covers such as the one on *BYTE* magazine in October 1993: 'State of the Art—How We'll Control Tomorrow's Computers'. But it proved to be an empty promise. Through press releases and partnerships with established manufacturers, GO had aroused the curiosity of the computer industry as a whole and encouraged a host

of others to attempt to produce a similar product, with varying levels of success, even before having a marketable product itself. The GO Computer even inspired another piece of deliberately misleading vapourware when Microsoft 'demonstrated' a non-existent forthcoming product in an attempt to detract media attention from GO. As it turned out, the technology could not be resolved adequately enough before all concerned lost interest, and the future direction of technology turned again. Even so, the widespread public dissemination of the concept of pen computing raised awareness among potential users and made them consider what computers could actually be like, arguably clearing the way for the ready acceptance of smaller, simpler pen-operated devices in the form of PDAs.

The previous points are by no means the only effects of vapourware, but the impact they had does at least prove a significant point. The focus of much of design history these days tends to downplay the production of design, choosing instead to concentrate on the consumption of mass-produced products, and great play is rightly made of the powerful forces of social construction. The study of vapourware, however, shows clearly

The first small production run of the Osborne 1 in an advertisement from 1981. It might have been inevitable someone would do it, but Osborne wasn't the first.

The Kaypro II, 1982. Directly influenced by the Osborne 1, itself a copy of Alan Kay's Xerox Notetaker.

that there remains some value in an area of study of products that never made it into manufacture, never appeared in the retail market, were never subjected to the capricious test of public opinion and yet still had significant effects on the development of computer history, our expectations of computers and our attitudes towards them. These products did not have to reach the market as finished products in order to have considerable effects in a number of ways. The mere concepts themselves were enough. Vapourware has not only raised awareness of current developments in the computer industry but has also led to completely new products that otherwise might never have appeared and which affected the path of computer history. Vapourware has at times acted in the same capacity as concept cars in the automobile industry: pointing the way forward for others, providing a target to be reached. In the process, it has enabled potential customers to come to terms with radical new ideas and unfamiliar forms of goods, making their acceptance easier when they do finally reach the market, often in watered-down form. Vapourware, it appears, can be a powerful force, and these pieces of vapourware are indeed proof, if any was ever needed, of the agency of ideas.

The Compaq Portable, 1983. The first legal IBM-compatible computer.

The Commodore SX-64, 1984. The first colour portable computer.

Timeline

- Difference Engine — 1821 to 1849
- Analytical Engine — 1834 to 1871
- 1939
- 1940
- Konrad Zuse Z3 computer built — 1941
- Atanasoff-Berry Computer (ABC) built — 1942
- Colossus first operated at Bletchley Park — 1943
- IBM ASCC (Harvard Mark 1) completed — 1944
- 1945
- ENIAC publicly announced by University of Pennsylvania — 1946
- The first transistor developed at Bell Labs — 1947
- IBM SSEC completed; Manchester 'Baby' (first stored program computer) trialled — 1948
- EDSAC built at Cambridge University; Manchester Mark 1 built at Manchester University — 1949
- Turing's Pilot ACE completed at National Physical Laboratory — 1950
- The first commercial computers sold: Ferranti Mark 1, LEO and UNIVAC I — 1951

Hofgaard Machine

Nordsieck Computer

KEY
Mainframe/Minicomputers Personal Computers Pen Computers Mobile Computers

- 1953 — IBM's first electronic computer, the IBM 701, built
- 1954 — Prototype transistor-based computer built at Manchester University
- 1956 — The transistorized computer the TX-0 built at MIT
- 1957 — Ferranti Pegasus built
- 1958 — First integrated circuit developed by Jack Kilby at Texas Instruments
- 1959 — DEC PDP-1
- 1963 — Doug Engelbart develops the computer mouse
- 1963 — Ivan Sutherland develops the 'Sketchpad' CAD software for the TX-2 computer
- 1964 — IBM introduces System/360, a family of compatible computers
- 1965 — Gordon Moore predicts computer power would double every year

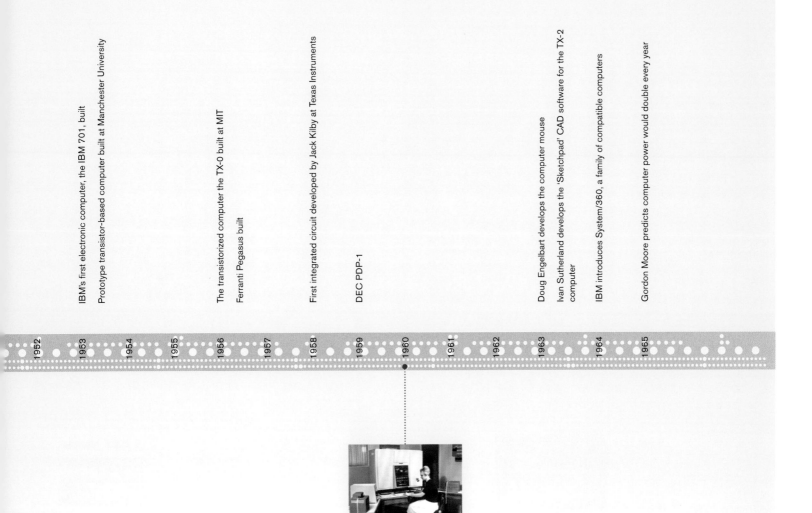

Saab D2

Timeline

- Mainframe computers using integrated circuits appear
- Doug Engelbart demonstrates 'Augment', the first interactive computer system
- Data General Nova launched
- UNIX developed at AT&T
- The Kenbak-1, an early personal computer, sold. Intel releases the first microprocessor
- The first email sent by Ray Tomlinson of Bolt, Beranek and Newman
- The computer arcade game Pong popularizes video games
- Xerox develops Alto, the first computer with a graphical user interface
- MITS sells the Altair 8800 as a kit; the Homebrew Computer Club meets in Menlo Park, CA
- IBM launches the IBM 5100 Portable Computer
- Apple 1 computer launched
- Cray 1 Supercomputer developed
- The Apple II, the TRS-80 and the Commodore PET create the market for home computers
- The IBM 5110 launched

1966 · 1967 · 1968 · 1969 · 1970 · 1971 · 1972 · 1973 · 1974 · 1975 · 1976 · 1977 · 1978

KEY
Mainframe/Minicomputers · Personal Computers · Pen Computers · Mobile Computers

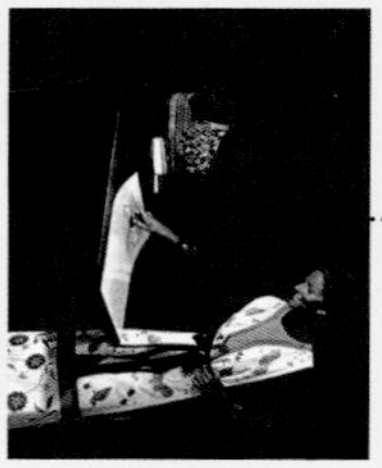

Honeywell Kitchen Computer

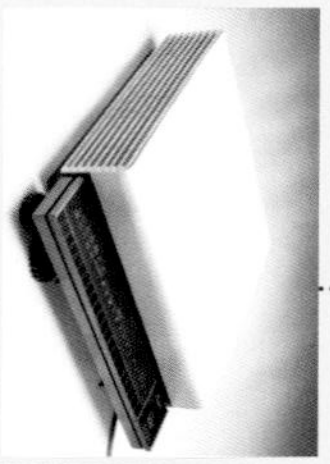

CTL Modular Three Minicomputer

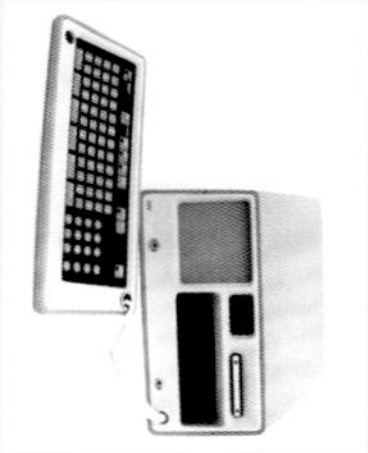

IBM Yellow Bird

Xerox Notetaker

Xerox Dynabook

IBM SCAMP Design Model

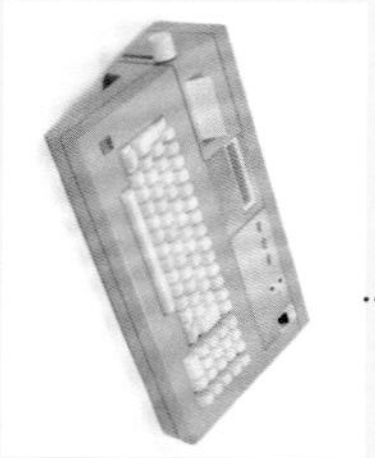

IBM Aquarius

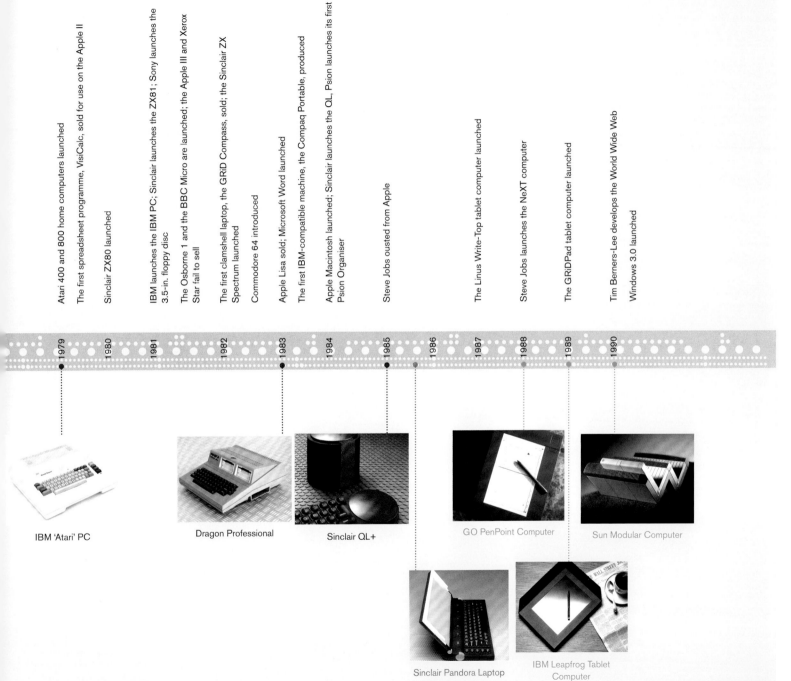

- Atari 400 and 800 home computers launched
- The first spreadsheet programme, VisiCalc, sold for use on the Apple II
- Sinclair ZX80 launched
- IBM launches the IBM PC; Sinclair launches the ZX81; Sony launches the 3.5-in. floppy disc
- The Osborne 1 and the BBC Micro are launched; the Apple III and Xerox Star fail to sell
- The first clamshell laptop, the GRiD Compass, sold; the Sinclair ZX Spectrum launched
- Commodore 64 introduced
- Apple Lisa sold; Microsoft Word launched
- The first IBM-compatible machine, the Compaq Portable, produced
- Apple Macintosh launched; Sinclair launches the QL, Psion launches its first Psion Organiser
- Steve Jobs ousted from Apple
- The Linus Write-Top tablet computer launched
- Steve Jobs launches the NeXT computer
- The GRiDPad tablet computer launched
- Tim Berners-Lee develops the World Wide Web
- Windows 3.0 launched

1979 · 1980 · 1981 · 1982 · 1983 · 1984 · 1985 · 1986 · 1987 · 1988 · 1989 · 1990

IBM 'Atari' PC

Dragon Professional

Sinclair QL+

GO PenPoint Computer

Sun Modular Computer

Sinclair Pandora Laptop

IBM Leapfrog Tablet Computer

Timeline

- 1991 — Linus Torvalds develops Linux
- 1992 — THE EO 440 communicator launched
- 1993 — The first Pentium processor developed
- 1993 — Apple sells the Newton MessagePad PDA
- 1994 — Netscape and Yahoo founded
- 1996 — Palm launches its first PDA, the Palm Pilot 1000
- 1997 — Steve Jobs returns to Apple
- 1998 — The Apple iMac is launched

1991 1992 1993 1994 1995 1996 1997 1998 1999 2000 2001 2002 2003 2004

Apple Figaro

EO Magni Personal Communicator

Phonebook

Siemens PIC

Compaq Dualworlds Notebook

Psion Halo and Ace

Pogo nVoy Communicator

KEY

Mainframe/Minicomputers Personal Computers Pen Computers Mobile Computers

Apple iPhone launched

Apple iPad launched

2005 2006 2007 2008 2009 2010

Palm Foleo

DualCor cPC

Acknowledgements

This book could not have been written without the help and support of a large number of people. My sincere apologies if I have missed anybody. In particular, I would like to thank the designers of many of the products that feature in the case studies for their time in giving interviews, answering queries and supplying images and other evidence, which often involved lengthy searching of their old archives: Jochen Backs, Paul Bradley, Tom Dair, Rick Dickinson, John Elliott, Nicola Guelfo, Tom Hardy, Jeff Hawkins, Marcus Hoggarth, Alan Kay, Don Kelemen, Kelly Kodama, Chris Loew, Peter H. Muller, Martin Riddiford and Morten Warren. A special, heartfelt thank you goes to the late Bill Moggridge for his initial introductions to so many of these people.

For their help in providing background information, pointing me in the right direction, finding images, giving permission to use images or otherwise providing support, I would like to thank Dag Andreassen, Alan Asarch, Marirosa Ballo, Celeste Baranski, John Barnes, Iann Barron, Alex Bochannek, Lasse Brunnström, Martin Campbell-Kelly, Stacy Castillo, Thomas Clifford, Ceinwen Cushway, Timo de Rijk, Benj Edwards, Celia Elliott, Rick English, Kjetil Fallan, Lee Felsenstein, Rob Ferris, Carmen Figini, Jim Fullalove, Tim Glass, Clive Grinyer, Simon Hardy, David Hembrow, Tom Jackson, Achim Jung, Jerry Kaplan, Barry Katz, Roel Klaassen, Erik Klein, Gordon Laing, Relsen Larsen, Paul Lasewicz, Simon Lavington, John Linney, David Linsley, Henry Madden, Susie Mulhern, Dick Nordsieck, Maria Rupert, Wendy Shay, Peter Skillman, Dag Spicer, Doron Swade, Kevin Symonds and Marc Weber.

I am indebted to Sheffield Hallam University, and in particular the Art and Design Research Centre, for their support, both financial and in kind. Thank you also to the staff at Bloomsbury: Tristan Palmer for his initial belief in the project, Simon Cowell for picking the project up, and Emily Ardizzone, Simon Longman and Noa Vazquez. Thanks to Emily Johnston at Apex CoVantage for her patience with requests to adjust layouts. For his sterling work on the preparation of images for print and providing great photographs, I have to thank Bernie Cavanagh (www.bernardcavanagh.com). I would also like to thank my family and friends for their continued support, with special thanks to Isaac for his patience while I wrote over so many weekends and to Sandra for her unending encouragement.

CALL FOR CASE STUDIES

If anyone reading this book has any suggestions for case studies of other computer vapourware projects suitable for inclusion in a second volume of *Delete*, please do not hesitate to get in touch with me via paul@paulatkinsondesign.com.

Picture Credits

Perpetual motion machine from G. A. Bockler's *Theatrum Machinarum Novum*, 1662. Science Museum Library/Science & Society Picture Library

Leonardo da Vinci's designs for a military tank and the 'air screw', an imagined helicopter, c. 1490. Science Museum/Science & Society Picture Library

The Futurama ride, part of the General Motors 'Highways and Horizons' exhibit at the World's Fair in New York, 1939. Courtesy of General Motors Media Archive

The Frigidaire 'Kitchen of the Future', 1956. Courtesy of General Motors Media Archive

Future transport predictions, 'Everyday Science and Mechanics', 1931, 1932.

General Motors GM–X Stiletto concept car, 1964. Courtesy of General Motors Media Archive

The AT&T Picturephone demonstrated at the World's Fair, 1964. Courtesy of AT&T Archives and History Center

Construction of the Difference Engine No. 2 built by the Science Museum in London, completed in 2002. Courtesy of Doron Swade

Early books of logarithmic tables, such as this example from 1670, were riddled with errors that regularly caused fatal shipwrecks. Science Museum/Science & Society Picture Library

Prototype of part of Babbage's Difference Engine No. 1, built by John Clements between 1824 and 1832. Science Museum/Science & Society Picture Library

Drawing by Charles Babbage of the side view of Difference Engine No. 2, late 1840s. Science Museum/Science & Society Picture Library

Poster for 'Making the Difference: Charles Babbage and the Birth of the Computer' exhibition, Science Museum, London, 1991. Courtesy of Doron Swade

Prototype of part of the Analytical Engine, 1834–1871. Science Museum/Science & Society Picture Library

Plan of the analytical engine drawn by Charles Babbage, c. 1840. Science Museum/Science & Society Picture Library

Punched cards for programming Babbage's Analytical Engine, 1834–1871. Science Museum/Science & Society Picture Library

Detail of the print mechanism of Babbage's Analytical Engine, 1834–1871. Science Museum/Science & Society Picture Library

The IBM ASCC (Automatic Sequence Controlled Calculator), 1944. Courtesy of IBM Corporate Archives

Demonstration model of Difference Engine No. 1 assembled from leftover components by Babbage's son Henry Prevost Babbage, c. 1880. Science Museum/Science & Society Picture Library

Prototype of the Hofgaard Machine with panels removed, c. 1955. Courtesy of Norwegian Museum of Science and Technology

A Francis Sønnichsen brochure for Lead-Acid batteries, c. 1950s. Courtesy of Norwegian Museum of Science and Technology

Fully assembled prototype of the Hofgaard Machine, c. 1955. Courtesy of Norwegian Museum of Science and Technology

Detail from the brochure of the Analyzer Corporation, c. 1960. Courtesy of Richard Nordsieck

The original Nordsieck computer, 1950. Courtesy of Department of Physics, University of Illinois at Urbana-Champaign

Arnold Nordsieck at the controls of his differential analyser, 1950. Courtesy of the (Champaign, IL) *News-Gazette*

Arnold Nordsieck with an early version of his gyroscope, c. 1954. Courtesy of Richard Nordsieck

Image from the brochure of the Analyzer Corporation, c. 1960. Courtesy of Richard Nordsieck

Working prototype of the Saab D2, 1960. Courtesy of Datamuseet IT-ceum

Hardware designer Gösta Neovius at the control panel of BARK, 1950. Courtesy of Datamuseet IT-ceum

Sweden's first electronic computer, BESK, c. 1956. Courtesy of Datamuseet IT-ceum

SARA, Saab's first binary computer system, 1957. Courtesy of Datamuseet IT-ceum

The Facit EDB, 1957. Courtesy of Datamuseet IT-ceum

The Saab D21 installed at the electric power company Skandinaviska Elverk in 1962. Courtesy of Datamuseet IT-ceum

The CK37 (Centralkalkylator) for the Saab 37 Viggen aircraft, 1971. Photo by Allan Näslund. Courtesy of Datamuseet IT-ceum

Image of the Kitchen Computer from *LIFE* magazine, 12 December 1969. © Yale Joel/Time & Life Pictures/Getty Images

Back cover of Honeywell's manual for the H316 minicomputer, 1969. Courtesy of Honeywell International Inc.

Don Kelemen's exploded drawing of the H316 minicomputer pedestal unit, 1969. Courtesy of Don Kelemen

Honeywell advertisement for the H316 minicomputer, Datamation, May 1969. Courtesy of Honeywell International Inc.

Neiman Marcus catalogue for the Kitchen Computer, 1969. © Science Photo Library

Data General Nova minicomputer, featured in *Datamation* magazine, November 1968

Neiman Marcus advertisement for the Kitchen Computer, 1969. Courtesy of the Computer History Museum

Prototype of the M3 minicomputer desktop unit by Bill Moggridge, 1973. Courtesy of IDEO

The Modular One installation at the Applied Psychology Unit (APU) in Cambridge, 1970. © MRC Cognition and Brain Sciences Unit, used by kind permission

The Modular One installation at the University of Birmingham being used in 1976. Paul Morby, University of Birmingham, 1976

Detail of the M3 'floating keyboard', 1973. Courtesy of IDEO

Early concept sketches for the M3 project, 1972. Courtesy of IDEO

An alternative desktop design for the M3 minicomputer by John Elliott, 1973. Courtesy of IDEO

Mock-up of the IBM SCAMP 'Design Model', 1973. Photo by Jim Casazza, used with permission of IBM Corporate Archives

The working 1973 prototype of the IBM SCAMP now at the Smithsonian Institution. Computer Collections, National Museum of American History, Smithsonian Institution.

The IBM SCAMP Design Model assembled for transit, 1973. Photo by Jim Casazza, used with permission of IBM Corporate Archives

The IBM 5100 Portable Computer, 1975. Courtesy of IBM Corporate Archives

The IBM 5110 Computer System, 1978. Courtesy of IBM Corporate Archives

The IBM 5150 PC, 1981. Courtesy of IBM Corporate Archives

Studio photograph of the IBM 'Yellow Bird' prototype, 1976. Photo by Jim Casazza, used with permission of IBM Corporate Archives

Prototype of the IBM 'Yellow Bird', 1976. Photo by Jim Casazza, used with permission of IBM Corporate Archives

Side view of the IBM 'Yellow Bird' with data cartridge inserted, 1976. Photo by Jim Casazza, used with permission of IBM Corporate Archives

Studio photograph of the IBM Aquarius in a domestic kitchen setting, 1977. Photo by Jim Casazza, used with permission of IBM Corporate Archives

The IBM Aquarius designed by Tom Hardy, 1977. Photo by Jim Casazza, used with permission of IBM Corporate Archives

Side view of the IBM Aquarius showing the slots for software cartridges and keypads. Photo by Jim Casazza, used with permission of IBM Corporate Archives

Plan view of the IBM Aquarius with software cartridge and keypad. Photo by Jim Casazza, used with permission of IBM Corporate Archives

A selection of solid-state software cartridges and function keypads for different IBM Aquarius programs. Photo by Jim Casazza, used with permission of IBM Corporate Archives

Working prototype of the Xerox Notetaker, 1978. © Mark Richards. Courtesy of the Computer History Museum

The state-of-the-art GRiD Compass, 1982. Courtesy of IDEO

The less capable but much cheaper Osborne 1, 1981. Courtesy of Steven Stengel, www.oldcomputers.net

The Xerox Alto computer, 1973. Courtesy of PARC, a Xerox company

Prototype of the IBM 'Atari' PC designed by Tom Hardy, 1979. Photo by Jim Casazza, used with permission of IBM Corporate Archives

The original Atari 800 Home Computer, designed by Kevin McKinsey, 1978. Courtesy of Benj Edwards

The original Atari 800 Home Computer, with cartridge cover raised, 1978. Courtesy of Benj Edwards

Sketches by Tom Hardy, analyzing the Atari mouldings for costing and considering how to 'convert' the form into an IBM product. Tom Hardy, used with permission of IBM Corporate Archives

Prototype of the QL+, 1985. Courtesy of Rick Dickinson

Sinclair QL, 1984. Image courtesy of Urs König, Sinclair QL Preservation Project http://tinyurl.com/sqpp25

The Sinclair Spectrum 128, 1985. Courtesy of Rick Dickinson

QL Wafer Expansion Module, 1985. Courtesy of Rick Dickinson

Prototype of the Super QL (rear view), 1986. Courtesy of Rick Dickinson

Concept drawing for the Mega PC Waferstack, 1986. Courtesy of Rick Dickinson

The GEC Dragon Professional, 1983. © Bernie Cavanagh

Packaging of the Dragon 32, 1982. © Bernie Cavanagh

Advertisement for the GEC-badged Dragon 64 in *Dragon User* magazine, July 1984. Courtesy of Simon Hardy, www.worldofdragon.org

The GEC Dragon Professional as featured in *Personal Computer World*, August 1984. Courtesy of Incisive Media

Cardboard mock-up of the Dynabook, 1968. Courtesy of PARC, a Xerox company

The 'Smalltalk' Graphical User Interface with overlapping windows. Courtesy of PARC, a Xerox company

The 'interim Dynabook system' being operated by children, 1973. Courtesy of PARC, a Xerox company

'MontBlanc' concept for the Apple Figaro project by Giugiaro Design, 1991. Courtesy of Giugiaro Design

First Figaro concept by Smart Design, 1989. Courtesy of Smart Design

First Figaro concept by Smart Design in transit, 1989. Courtesy of Smart Design

Second Figaro concept by Smart Design, 1989. Courtesy of Smart Design

Second Figaro concept by Smart Design in situ, 1989. Courtesy of Smart Design

First Figaro concepts from Giugiaro Design, 1989. Courtesy of Giugiaro Design

First Figaro concepts from Giugiaro Design, 1989. Courtesy of Giugiaro Design

Revised Apple Figaro concepts by Giugiaro Design, 1990. Courtesy of Giugiaro Design

Apple Figaro 'red eye' concept by Giugiaro Design, 1990. Courtesy of Rick English

The Apple Newton MessagePad 100, 1993. Courtesy of Rick English

Sun modular computer central processing unit concept, 1990. Courtesy of Paul Bradley

Sun modular computer central processing unit concept in upright position. Courtesy of Paul Bradley

Self-contained modular pen computer concept, 1990. Courtesy of Paul Bradley

Prototype of the GO PenPoint Computer with external accessories, 1991. Courtesy of Paul Bradley

Prototype of the GO PenPoint Computer and PenPoint interface, 1991. Courtesy of Paul Bradley

One of the sixty-five demonstration machines on the cover of *Personal Computer World*, April 1991. Courtesy of Incisive Media

Examples of the gestures used to enter commands in the PenPoint interface. Courtesy of Incisive Media

Working prototype of the IBM Leapfrog Tablet Computer, 1992. Photo by Sally Andersen-Bruce, used with permission of IBM Corporate Archives

Cover of *I.D.* magazine, May–June 1993. Courtesy of F+W Media

The IBM Leapfrog Tablet Computer on its 'lilypad' base unit. Copyright Aldo Ballo, originally appeared in Domus February 1994 /Editoriale Domus S.p.A.

Model of the AT&T EO Magni Personal Communicator, 1993–1994. Courtesy of Paul Bradley

Frog design's advertisement featuring the EO Personal Communicator, *Design* magazine, June 1993. Courtesy of Paul Bradley

EO promotional brochure, 1992. Courtesy of David Hembrow

Blue sky 'personal valet' concept for EO by Naoto Fukusawa, 1993. Courtesy of Paul Bradley

Model of the AT&T EO Magni Personal Communicator, 1993–1994. Courtesy of Paul Bradley

Clear acrylic prototype of Project 'Loki', 1994. Courtesy of Henry Madden

Publicity image of the DualCor cPC, 2006. Courtesy of DualCor Technologies Inc.

Kelly Kodama's design for the Chameleon, 2001. Courtesy of Zoe Design Associates

Appearance model of the Chameleon, 2001. Courtesy of Zoe Design Associates

Chris Loew's redesign of the cPC, 2005. Courtesy of Loewco

The 'engineering sample' of the DualCor cPC with wireless modem, 2006. Courtesy of Peter H. Muller, Interform

Proposed redesigns of cPC by Peter H. Muller, 2006. Courtesy of Peter H. Muller, Interform

Prototype of the Pandora Laptop, 1985. Courtesy of Rick Dickinson

Sinclair ZX80, 1980. Courtesy of Rick Dickinson

Sinclair ZX81, 1981. Courtesy of Rick Dickinson

ZX Spectrum and Microdrives, 1983. Courtesy of Rick Dickinson

TV80 plus development models, 1984. Courtesy of Rick Dickinson

Models of different versions of the Pandora Laptop, 1985–1986. Courtesy of Rick Dickinson

Cambridge Computers Z88, 1988. Courtesy of Rick Dickinson

Working prototype of the Phonebook, 2003. Courtesy of Therefore Product Design Consultants

The Psion Series 3a, 1993. © Bernie Cavanagh

'Thor', an early version of the Phonebook concept under the Psion brand, 1999. Courtesy of Therefore Product Design Consultants

Phonebook variants branded for Vodafone and T-Mobile, 2003. Courtesy of Therefore Product Design Consultants

Phonebook prototype closed and being opened, 2003. Courtesy of Therefore Product Design Consultants

'Oyster' Phonebook for T-Mobile, 2004. The result of design by committee. Courtesy of Therefore Product Design Consultants

Vodafone 'Fonebook' branded version, 2004. Courtesy of Therefore Product Design Consultants

The Psion Organiser II, 1986. Courtesy of Therefore Product Design Consultants

The Psion Series 5, 1997. Courtesy of Therefore Product Design Consultants

3D CAD of the final PIC design used to produce soft tooling, 1999. Courtesy of Therefore Product Design Consultants

Studio photographs of the Siemens PIC, 1999. Courtesy of Therefore Product Design Consultants

Psion Ace mobile phone/PDA and Halo headset concepts, 1999. Courtesy of Therefore Product Design Consultants

Concept model of the Psion Ace and earpiece, closed and partially opened. Courtesy of Therefore Product Design Consultants

Concept model of the Psion Ace fully open. Courtesy of Therefore Product Design Consultants

Concept model of the Psion Halo, 1999. Courtesy of Therefore Product Design Consultants

An image of the Psion Halo projecting onto a user's hand. Courtesy of Therefore Product Design Consultants

Compaq Dualworlds notebook prototype designed by Morten Warren, 2001. Courtesy of Native Design Ltd

Alternative design for the Dualworlds notebook, 2001. Courtesy of Native Design Ltd

Alternative designs for the Dualworlds notebook, 2001. Courtesy of Native Design Ltd

Compaq Dualworlds notebook, closed for carrying, 2001. Courtesy of Native Design Ltd

Compaq Dualworlds notebook, open in notebook format, 2001. Courtesy of Native Design Ltd

Side view of the Compaq Dualworlds notebook in desktop format, 2001. Courtesy of Native Design Ltd

Prototype of the Pogo nVoy Communicator, 2003. Courtesy of Pogo Mobile Solutions

Concept sketches of the Pogo by Marcus Hoggarth while at Therefore, 2000. Courtesy of Pogo Mobile Solutions

The original Pogo, 2001. Courtesy of Pogo Mobile Solutions

The original Pogo in use, 2001. Courtesy of Pogo Mobile Solutions

Exploded view of the nVoy e100 prototype, 2003. Courtesy of Pogo Mobile Solutions

Working prototype of the nVoy e100, 2003. Courtesy of Pogo Mobile Solutions

Final working version of the Palm Foleo mobile companion, 2006. Courtesy of Businesswire

Initial concept sketch of 'Hollywood', 2003. Courtesy of Peter Skillman

Final working version of the Palm Foleo mobile companion open and closed, 2006. Photos by Paul Atkinson

The Scheutzian Calculation Engine, 1853. Directly inspired by Babbage's well-publicized work. Science Museum/Science & Society Picture Library

The first small production run of the Osborne 1 in an advertisement from 1981. It might have been inevitable someone would do it, but Osborne wasn't the first. Courtesy of David Scott Carlick

The Kaypro II, 1982. Directly influenced by the Osborne 1, itself a copy of Alan Kay's Xerox Notetaker. Courtesy of Steven Stengel, www.oldcomputers.net

The Compaq Portable, 1983. The first legal IBM-compatible computer. Courtesy of Steven Stengel, www.oldcomputers.net

The Commodore SX-64, 1984. The first colour portable computer. Courtesy of Steven Stengel, www.oldcomputers.net

Notes

IMAGINED MACHINES

1. Roger Bacon, cited in Marco Cianchi, *Leonardo da Vinci's Machines* (Becocci Editore, 1988), p. 12.
2. George Basalla, *The Evolution of Technology* (Cambridge University Press, 1988), p. 68.
3. Cianchi, *Leonardo da Vinci's Machines*, p. 82.
4. Ibid.
5. Brian Ash, ed., *The Visual Encyclopedia of Science Fiction* (Book Club Associates, 1978), passim.
6. Basalla, *Evolution of Technology*, p. 77.
7. Mike Ashley, *Out of This World: Science Fiction but Not As You Know It* (The British Library, 2011), p. 6.
8. Joseph J. Corn and Brian Horrigan, *Yesterday's Tomorrows: Past Visions of the American Future* (Johns Hopkins University Press, 1984), p. xiii.
9. Richard Wurts, *The New York World's Fair, 1939/1940* (Dover Publications, 1977), p. 20.
10. David Gelernter, *1939, The Lost World of the Fair* (The Free Press, 1995), p. 22–3.
11. Nicolas P. Maffei, '"I Have Seen the Future": Norman Bel Geddes' "Futurama" as Immersive Design', *Design and Culture* 4/1, p. 80.
12. D. Spicer, 'If You Can't Stand the Coding, Stay Out of the Kitchen', based on 'The Archeology of Computing', delivered at the conference Alien Intelligence, Kiasma, the National Museum of Contemporary Art, Helsinki, Finland, March 2000; published in *Dr. Dobb's Journal* [online journal], 12 August 2000, http://www.ddj.com/architect/184404040, accessed 31 May 2012.
13. Narration in *1999 A.D.* (Philco-Ford Corporation, 1967).
14. Corn and Horrigan, *Yesterday's Tomorrows*, p. xiii.
15. Ibid., p. 100.
16. J. Pettifer and N. Turner, *Automania: Man and the Motor Car* (Guild Publishing, 1984), p. 133.
17. Ashley, *Out of This World*, p. 6.
18. Ibid., p. 87.
19. Kenneth Lipartito, 'Picturephone and the Information Age: The Social Meaning of Failure', *Technology and Culture* 44/1 (2003), pp. 51–2.
20. Hans-Joachim Braun, 'Symposium on Failed Innovations: Introduction', *Social Studies of Science* 22/2 (1992), p. 213.
21. T. J. Pinch and W. E. Bijker, 'The Social Construction of Facts and Artifacts: Or How the Sociology of Science and the Sociology of Technology Might Benefit Each Other', in W. E. Bijker, T. P. Hughes and T. J. Pinch, eds., *The Social Construction of Technological Systems: New Directions in the Sociology and History of Technology* (MIT Press, 1987), p. 40.
22. Lipartito, 'Picturephone and the Information Age', p. 53.
23. Ibid., p. 52.
24. P. A. David, 'Clio and the Economics of QWERTY', *American Economic Review* 75/2 (1985), p. 332.
25. Ibid., p. 335.
26. A painting by the 'technological futurist' Syd Mead of a huge flat-screen 3D projection television executed in the early 1970s while he worked at Philips CIDC in Eindhoven appears in Syd Mead, *Sentinel* (Dragon's Dream, 1978).
27. Nicky Trevett, 'Television Sets', *Design Magazine* (April 1994), p. 47.
28. Dennis Normile, 'Reinventing the Cathode Ray Tube', *Popular Science* (February 1994), p. 34.
29. The phrase is attributed to the French poet Paul Valéry in the mid 1940s and has been used by many others. Arthur C. Clarke used it in a way perhaps most relevant to this discussion in an article predicting life in 2001: Arthur C. Clarke, 'The Future Isn't What It Used to Be', *Engineering and Science* 33/7 (1970), pp. 4–9.
30. Jake van Slatt, 'A Steampunk Manifesto', in Jeff VanderMeer, *The Steampunk Bible: An Illustrated Guide to the World of Imaginary Airships, Corsets, Goggles, Mad Scientists, and Strange Literature* (Abrams Image, 2011), p. 216.

MAINFRAMES AND MINICOMPUTERS

1. IBM, 'Herman Hollerith', *IBM Archives*. http://www-03.ibm.com/ibm/history/exhibits/builders/builders_hollerith.html, accessed 16 April 2012.
2. Frank Carter, 'The Turing Bombe', *The Rutherford Journal* 3 (2010). http://www.rutherfordjournal.org/article030108.html, accessed 16 April 2012.
3. IBM, 'IBM's ASCC', *IBM Archives*. http://www-03.ibm.com/ibm/history/exhibits/markI/markI_intro.html.
4. W. O. Baker, 'The First Ten Years of the Transistor', *The Bell System Technical Journal* 37/5 (September 1958), pp. i–vi.
5. IBM, 'System/360 Announcement', *IBM Archives*. http://www-03.ibm.com/ibm/history/exhibits/mainframe/mainframe_PR360.html, accessed 16 April 2012.
6. IBM, 'Mainframe Concepts', IBM.com. http://publib.boulder.ibm.com/infocenter/zos/basics/index.jsp?topic=/com.ibm.zos.zmainframe/zconc_whatismainframe.htm, accessed 16 April 2012.
7. Doron Swade, 'The Shocking Truth about Babbage and His Calculating Engines', *Resurrection: The Bulletin of the Computer Conservation Society* 32 (2004). http://www.cs.man.ac.uk/CCS/res/res32.htm#d, accessed 9 November 2010.
8. Doron Swade, 'Charles Babbage's Difference Engine No. 2 Technical Description', *Science Museum Papers in the History of Technology* 4 (1995).
9. Doron Swade, 'Automatic Computation: Charles Babbage and Computational Method', *The Rutherford Journal* 3 (2010). http://www.rutherfordjournal.org/article030106.html, accessed 9 November 2010. The Alan Turing paper referred to is the seminal 'On Computable Numbers with an Application to the Entscheidungs problem', written in 1936, and published in the *Proceedings of the London Mathematics Society* s2-42/1 (1937), pp. 230–65.
10. Swade, 'Automatic Computation'.
11. Swade, 'Shocking Truth about Babbage'.
12. Swade, 'Automatic Computation'.
13. Doron Swade, 'Redeeming Charles Babbage's Mechanical Computer', *Scientific American* (February 1993), p. 63.
14. Ibid.
15. Louise Purbrick, 'The Dream Machine: Charles Babbage and His Imaginary Computers', *Journal of Design History* 6/1 (1993), p. 12.
16. Swade, 'Redeeming Charles Babbage's Mechanical Computer', p. 65.
17. Swade, 'The Construction of Charles Babbage's Difference Engine No. 2', *IEEE Annals of the History of Computing* (July–September 2005), p. 70. This paper provides a detailed account of the effort to construct the Difference Engine No. 2 by the team at the Science Museum.
18. Swade, 'Automatic Computation'.
19. The exhibition, 'Making the Difference: Charles Babbage and the Birth of the Computer', opened at the Science Museum, London, June 1991.
20. Swade, 'Redeeming Charles Babbage's Mechanical Computer', p. 67.
21. Doron Swade, personal communication with the author, 7 December 2010.
22. Swade, 'Automatic Computation'.
23. John von Neumann described this architecture in his 'First Draft of a Report on the EDVAC', a set of notes synthesized from meetings he attended at the Moore School of Electrical Engineering at the University of Pennsylvania, published and distributed in June 1945.
24. Doron Swade, email communication with the author, 8 November 2010.
25. Purbrick, 'Dream Machine', p. 12.
26. I. Bernard Cohen, 'Babbage and Aiken', *Annals of the History of Computing* 10/3 (1988), p. 175.
27. Ibid., p. 174.
28. Ibid., p. 183.
29. Ibid., p. 172.
30. '100 Years as Technology Leaders', 17 December 2009. http://www.proxll.no/index.php?option=com_content&view=article&id=296&Itemid=307&lang=en, accessed 4 November 2011.
31. 'Hofgaardmaskinen: Et FOU-prosjekt som preget Sønnicos historie' (The Hofgaard Machine: an R&D project that characterized Sønnico's history). Sønnico Annual Report, 1986, p. 14. Translated by the author. Original in the Norwegian Museum of Science, Technology and Medicine.
32. Ibid.
33. Ibid., p. 15.

34 Ibid.

35 '100 Years as Technology Leaders'.

36 Richard Nordsieck, email communication with the author, 23 January 2012.

37 Arnold Nordsieck, 'The Nordsieck Computer', *Proceedings of the Western Computer Conference—Joint IRE-AIEE-ACM* (4–6 February 1953), p. 227.

38 'U.I. Board Releases Patent on Calculator', *News-Gazette*, 26 November 1951.

39 Nordsieck email, 7 December 2011.

40 'Illini Scientist Builds "Brain" with $700', *News-Gazette*, 2 April 1950.

41 Nordsieck, 'The Nordsieck Computer', p. 230.

42 'Illini Scientist Builds "Brain"'.

43 Charlie Blue, 'A Very Modern Device', *Lawrence Livermore National Laboratory 50th Anniversary Newsletter*, 2002, p. 9.

44 Nordsieck email, 7 December 2011.

45 Ibid.

46 For details, see the Institute of Navigation website at http://www.ion.org/museum/item_view.cfm?cid=2&scid=4&iid=30, accessed 30 January 2012.

47 Tord Jöran Hallberg, *IT Gryning: Svensk datahistoria från 1840- till 1960-talet* (Lund, Sweden: Studentlitteratur, 2007), p. 373.

48 Ibid., p. 135.

49 Ibid., p. 374.

50 Ibid., p. 159.

51 Beaten only by the American ERA 1103 from Engineering Research Associates (Hallberg, *IT Gryning*, p. 374).

52 The Sunet Archive, Swedish University Computer Network. http://ftp.sunet.se/pub/pictures/computer/SMIL-00.README, accessed 1 November 2011.

53 Thomas Clifford, curator, Datamuseet, Linköping, personal communication with the author, 6 October 2011.

54 Hallberg, *IT Gryning*, p. 377.

55 Ibid., p. 380.

56 Stanley Marcus, *Minding the Store* (Elm Tree Books, 1975), p. 229.

57 Gordon Bell, *A Congeries on the Computer-in-the-home Market*, Internal Memorandum of Digital Equipment Corporation, 11 December 1969. Accession no. 102630372, Archives of the Computer History Museum.

58 Paul Atkinson, 'The Curious Case of the Kitchen Computer: Products and Non-products in Design History', *Journal of Design History* 23/2, p. 173.

59 Don Kelemen, email communication with the author, 27 May 2009.

60 Martin Pawley, 'Office beneath the Skin', *Design* (March 1970), p. 18.

61 APU Timeline. http://www.mrc-cbu.cam.ac.uk/history/timeline/timeline.html, accessed 12 December 2011.

62 'The APU's First Proper Computer (1970)', Applied Psychology Unit, Cambridge. http://www.mrc-cbu.cam.ac.uk/history/electronicarchive/modular1.html, accessed 12 December 2011.

63 'The History of Computer Science', School of Computer Science, University of Birmingham. http://www.cs.bham.ac.uk/about/history/history1970.php, accessed 12 December 2011.

64 Iann Barron, email communication with the author, 27 February 2012.

65 Bill Moggridge, email communication with the author, 22 January 2009.

66 John Elliott, personal communication with the author, 27 February 2012.

67 Barron email, 27 February 2012.

68 Mark Brutton, 'Just for the Look of the Thing', *Design* 368 (August 1979), p. 59.

69 Barron email, 27 February 2012.

PERSONAL AND PORTABLE COMPUTERS

1 Douglas Engelbart, the inventor of the computer mouse, for one, predicted in the 1960s that individuals would own their own interactive computers, but he recalled that whenever he mentioned it, it was like saying that everyone would one day own their own helicopter. Douglas Engelbart, interview with the author, 10 April 2006.

2 Gordon Moore first published his findings that became commonly known as 'Moore's Law' in a 1965 issue of *Electronics* magazine.

3 Early home computers were 'self-referential' machines, really only useful for learning about computers themselves. See Leslie Haddon, 'The Home Computer: The Making of a Consumer Electronic', *Science as Culture* 2 (1988), p. 27.

4 Paul Atkinson, 'Actor Networks and the Development of the Home Computer', in Fiona Hackney, Jonathan Glynne and Viv Minton, eds., *Networks of Design* (Universal Publishers, 2009), p. 223.

5 Apple finally achieved this accolade in 2011. See Dominic Rushe, 'Apple Pips Exxon as World's Biggest Company', *The Guardian*, 10 August 2011, p. 24.

6 Owen Linzmayer, *Apple Confidential 2.0: The Definitive History of the World's Most Colorful Company* (No Starch Press, 2008), p. 5.

7 'IBM Personal Computer', IBM Archives. http://www-03.ibm.com/ibm/history/exhibits/pc/pc_1.html, accessed .

8 Jonathan Littmann, 'The First Portable Computer: The genesis of SCAMP, grandfather of the personal computer', *PC World*, October 1983, p. 296. Note that this project is not to be confused with an earlier 48-bit mainframe project with the same name, which was developed by John Fairclough at the IBM World Trade Corporation's Hursley Laboratory in the UK between 1958 and 1961. See William Rodgers, *Think*, London, Weidenfeld & Nicolson, 1969, p. 287, and the memoirs of Bob Evans, *The Genesis of the Mainframe*, available online at http://researcher.watson.ibm.com/researcher/files/us-bbfinkel/bob_o_evans_mainframe.pdf. Accessed 29 October 2012.

9 Tom Hardy, interview with the author at Savannah College of Art and Design, 17 February 2011.

10 Littmann, 'First Portable Computer', p. 296.

11 The 'A' in the SCAMP acronym stands for 'APL'—itself an acronym of a computer programming language, which took its initials from a 1962 book describing the system, *A Programming Language*, by its inventor, Kenneth E. Iverson.

12 IBM 5100 Portable Computer brochure, 1976.

13 Littmann, 'First Portable Computer', p. 297.

14 Ibid.

15 Ibid., p. 298.

16 Ibid., p. 299.

17 Hardy interview, 17 February 2011.

18 Ibid.

19 'IBM 5100 Portable Computer', IBM Archives. http://www-03.ibm.com/ibm/history/exhibits/pc/pc_2.html, accessed 28 November 2011.

20 'IBM 5100 Portable Computer', IBM Archives. http://www-03.ibm.com/ibm/history/exhibits/pc/pc_3.html, accessed 28 November 2011.

21 'IBM 5110', IBM Archives. http://www-03.ibm.com/ibm/history/exhibits/pc/pc_4.html, accessed 28 November 2011.

22 'IBM System/23 Datamaster', IBM Archives. http://www-03.ibm.com/ibm/history/exhibits/pc/pc_9.html, accessed 29 November 2011.

23 Littmann, 'First Portable Computer', p. 300.

24 Hardy interview, 17 February 2011.

25 Vannevar Bush, in an article titled 'As We May Think', *Atlantic Monthly* (July 1945), described an individual machine called the 'Memex', which would store all of a user's information. This inspired a number of people, including the inventor of the computer mouse, Douglas Engelbart, to develop personal computer systems.

26 Atkinson, 'Actor Networks', pp. 219–24.

27 Hardy interview, 17 February 2011.

28 Ibid.

29 See US Patent US4007453 (A). http://worldwide.espacenet.com/publicationDetails/biblio?FT=D&date=19770208&DB=EPODOC&locale=en_EP&CC=US&NR=4007453A&KC=A&ND=4, accessed 13 April 2012.

30 As it happens, bubble memory never did take off. It proved to have some technical reliability problems of its own, and as disk drives developed in the 1980s, they became more reliable, cheaper and capable of holding more information.

31 Hardy interview, 17 February 2011.

32 Ibid.

33 Ibid.

34 'Computers: Carry Along, Punch In, Read Out', *Time* (21 June 1982). http://www.time.com/time/printout/0,8816,925484,00.html, accessed 4 December 2011.

35 I. Stobie, 'They All Laughed, But…', *Practical Computing* (January 1983), p. 108.

36 Michael Hiltzik, *Dealers of Lightning* (Orion Business, 2000), p. 318.

37 Ibid., p. 321.

38 Ibid., p. 327.

39 Ibid., p. 328.

40 Ibid., p. xvii.

41 Alan Kay, email communication with the author, 31 August 2008.

42 Hiltzik, *Dealers of Lightning*, p. 327.

43 Hardy interview, 17 February 2011.

44 Ibid.

45 Ibid.

46 'The Birth of the IBM PC', *IBM Archives*. http://www-03.ibm.com/ibm/history/exhibits/pc25/pc25_birth.html, accessed 4 April 2012.

47 Hardy interview, 17 February 2011.

48 Benj Edwards, 'Atari's Answer to the Apple II', *PC World.com* (2009). http://www.pcworld.com/article/181421/inside_the_atari_800.html, accessed 3 April 2012.

49 'Retro Home Computers—Atari 800'. http://retrogameandcomputer.com/atari-800-retro-computer.php, accessed 3 April 2012.

50 Hardy interview, 17 February 2011.

51 Ibid.

52 Ibid.

53 'The Birth of the IBM PC'.

54 Rick Dickinson, email communication with the author, 12 April 2011.

55 'Sinclair Exploits Old Technology for New Chip', *New Scientist* (27 September 1984), p. 27.

56 Ian Adamson and Richard Kennedy, *Sinclair and the 'Sunrise Technology'* (Penguin Books, 1986), p. 176.

57 Rick Dickinson, interview with the author, 19 November 2010.

58 Brendon G., 'Editorial', *Dragon User* (September 1984), p. 3.

59 'Another String to the Past Cut', *Dragon User* (December 1983), p. 11.

60 'No Dragons at Smiths', *Dragon User* (May 1983), p. 7.

61 Bain K. and Bain S., 'The 64: How It Rates', *Dragon User* (December 1983), p. 22.

62 'More Micros to Follow Memory', *Dragon User* (May 1983), p. 7.

63 Cunningham, G., 'New Man in the Driver's Seat', *Dragon User* (December 1983), p. 34.

64 Hannah C., 'Dragon 64 Gets Ready for the US', *Dragon User* (September 1983), p. 19.

65 John Linney, email communication with the author, 9 August 2011.

66 GEC had previously tried to enter the home computer market through British computer manufacturers Torch Computers Ltd (Cunningham, G., 'Editorial', *Dragon User* (October 1983), p. 3.)

67 Ibid.

68 'Another String to the Past Cut', p. 11.

69 Cunningham, 'New Man in the Driver's Seat', p. 34.

70 L. Coley, 'Dragon Professional', *Personal Computer World* (August 1984).

71 From discussions with Simon Hardy and John Linney. The early prototypes had a top 'moulding' fabricated from acrylic, and these did not have serial numbers. The injection moulded prototypes were sent to reviewers and shown at Earl's Court. The 'Earl's Court' model owned by Simon Hardy is shown in the main photograph and has the serial number 000008, suggesting at least seven others were produced.

72 'The Future of Dragon', *Dragon User* (August 1984), p. 8.

PEN COMPUTING

1 See, for example, M. Fisher, 'Momenta Head to Offer His "Pentop" Computer', *New York Times*, 5 October 1991.

2 B. Breen, 'Fresh Start 2002: Starting Over … and Over…', *Fast Company* 54 (December 2001), p. 77.

3 Jeff Hawkins, email communication with the author, 24 January 2007.

4 C. H. Blickensdorfer, '10 Years of Pen Computing', *Pen Computing Magazine* 50 (June 2004).

5 http://research.microsoft.com/en-us/um/redmond/events/fs2005/presentations/FacultySummit_2005_Keely.ppt, accessed 22 November 2011.

6 Alan Kay and Adele Goldberg, 'Personal Dynamic Media,' *Computing* 10/3 (March 1977), p. 31.

7 Alan Kay, 'The Early History of Smalltalk', *ACM SIGPLAN Notices* 28/3 (March 1993), p. 71.

8 Alan Kay, 'FLEX: A Flexible Extendable Language', MSc thesis, University of Utah, 1968.

9 Kay, 'Early History of Smalltalk', p. 72.

10 Ibid., p. 73.

11 Alan Kay, email communication with the author, 8 August 2007.

12 Kay, 'Early History of Smalltalk', p. 73.

13 Kay email, 8 August 2007.

14 In early accounts by Kay, this weight appears as 4 lbs. In later accounts it is stated as 2 lbs.

15 Alan Kay, 'The Reactive Engine', PhD thesis, University of Utah, 1969.

16 Kay's famous quote 'The best way to predict the future is to invent it' was said in 1971 to Xerox manager Don Pendery in response to his directive to think how Xerox could defend itself against future competition. Kay, 'Early History of Smalltalk', p. 75.

17 Ibid., p. 76.

18 The first of these public papers was Alan Kay, 'A Personal Computer for Children of All Ages', *Proceedings of the ACM National Conference, Boston* (August 1972). It contains a narrative story of two children using Dynabooks in all aspects of their play and study, illustrated with the cartoon he drew in 1968.

19 Kay, 'Early History of Smalltalk', p. 76.

20 Michael Hiltzik, *Dealers of Lightning* (Orion Business, 2000), p. 264–5.

21 John Lees, 'The World in Your Own Notebook', in David Ahl and Burchenal Green, eds., *The Best of Creative Computing*, iii (Creative Computing Press, 1980), p. 6.

22 Squeakland, 'Alan Kay'. http://www.squeakland.org/about/people/bio.jsp?Id=1.

23 Paul Kunkel, *Apple Design: The Work of the Apple Industrial Design Group* (Graphis US Inc., 1997), p. 74.

24 Tom Dair, 'Smart Design's iPad … Circa 1989', *Fast Company* (26 March 2010). http://www.fastcompany.com/1598501/smart-designs-ipad-circa-1989, accessed 24 April 2012.

25 Tom Dair, personal communication with the author, 24 April 2012.

26 Dair, 'Smart Design's iPad … Circa 1989'.

27 Kunkel, *Apple Design*, p. 76.

28 Ibid., p. 84.

29 See Rishab Aiyer Ghosh, *Study on the Economic Impact of Open Source Software on Innovation and the Competitiveness of the Information and Communication Technologies (ICT) Sector in the EU*, European Commission report 0.5754, 2006. http://ec.europa.eu, accessed 16 March 2012.

30. Stanford University, 'Wellspring of Innovation: Sun Microsystems Spotlight'. http://www.stanford.edu/group/wellspring/sun_spotlight.html, accessed 16 March 2012.
31. Funding Universe, 'Sun Microsystems Inc., Company History'. http://www.fundinguniverse.com/company-histories/Sun-Microsystems-Inc-company-History.html, accessed 13 March 2012.
32. Paul Bradley, interview with the author at the offices of frog design, San Francisco, 15 February 2011.
33. Jochen Backs, email communication with the author, 21 May 2012.
34. Bradley interview, 15 February 2011.
35. See Jack Schofield, 'Larry Ellison Starts to Reign over Sun', *The Guardian Technology Blog*, 13 May 2010. http://www.guardian.co.uk/technology/blog/2010/may/13/ellison-oracle-sun/print, accessed 12 March 2012.
36. Charles Arthur, 'Jonathan Schwartz Tweets His Last Goodbye to Sun Microsystems', *The Guardian Technology Blog*, 4 February 2010. http://www.guardian.co.uk/technology/blog/2010/feb/04/jonathan-schwartz-sun-microsystems-tweet-ceo-resignation, accessed 12 March 2012.
37. Jerry Kaplan, *Start Up: A Silicon Valley Adventure* (Little, Brown and Co., 1994), p. 15.
38. Ibid., p. 9.
39. Kaplan, *Start Up* (Houghton Mifflin Company, 1994). Edition quoted here published by Little, Brown and Co., 1995.
40. Ibid., p. 56.
41. Ibid., p. 62.
42. Ibid., p. 106.
43. Jennifer Edstrom and Marlin Eller, *Barbarians Led by Bill Gates: Microsoft from the Inside* (H. Holt, 1998), p. 120.
44. Celeste Baranski, email communication with the author, 13 October 2011.
45. Paul Bradley, interview with the author at the offices of IDEO, Palo Alto, 8 May 2007.
46. Rupert Goodwins, 'Go PenPoint', *Personal Computer World* (April 1991), p. 140.
47. Baranski email, 13 October 2011. In *Start Up* (p. 111), Kaplan quoted Baranski as saying, 'The battery should last about an hour'.
48. Kaplan, *Start Up*, p. 110.
49. Ibid., p. 142.
50. Ibid., p. 148.
51. Ibid., p. 160.
52. Ibid., p. 164.
53. Edstrom and Eller, *Barbarians Led by Bill Gates*, p. 130.
54. Ibid., p. 138.
55. Kaplan, *Start Up*, p. 171.
56. Cover of *Byte* magazine, February 1991.
57. Cover of *Personal Computer World*, April 1991.
58. Kaplan, *Start Up*, p. 202.
59. Ibid., p. 294.
60. Kiyonori Sakakibara, 'Global New Product Development: The Case of IBM Notebook Computers', *Business Strategy Review* 6/2 (1995), p. 36.
61. Julie Trelstad, 'Code Name Leapfrog', *I.D.: The International Design Magazine* (May/June 1993), p. 71.
62. Sakakibara, 'Global New Product Development', p. 26.
63. Tom Hardy, 'Innovation and Chaos', *Design Management Journal* 5/3 (Summer 1994), p. 38.
64. Trelstad, 'Code Name Leapfrog', p. 71.
65. Ibid.
66. Hardy, 'Innovation and Chaos', p. 38.
67. Trelstad, 'Code Name Leapfrog', p. 71.
68. Marco Romanelli, 'Leapfrog Computer', *Domus* (February 1994), p. 64.
69. The name ThinkPad became a generic name applied to all IBM laptops, but the original IBM ThinkPad—the 700T—was a pen-based tablet launched in 1992. See video at http://hothardware.com/News/Lenovo-Explains-First-ThinkPad-Tablet-On-Video-Promises-New-Products/, accessed 1 December 2011.
70. Richard Sapper, quoted in Romanelli, 'Leapfrog Computer', p. 66.
71. Tom Hardy, interview with the author at Savannah College of Art and Design, 17 February 2011.
72. Trelstad, 'Code Name Leapfrog', p. 73.
73. Hardy, 'Innovation and Chaos', p. 38.
74. Trelstad, 'Code Name Leapfrog', p. 73.
75. Hardy, 'Innovation and Chaos', p. 38.
76. Trelstad, 'Code Name Leapfrog', p. 73.
77. Ibid., p. 71.
78. Hardy interview, 17 February 2011.
79. Trelstad, 'Code Name Leapfrog', p. 71.
80. Hardy interview, 17 February 2011.
81. Hardy, 'Innovation and Chaos', p. 39.
82. Kaplan, *Start Up*, p. 227. The first Apple Newton actually shipped in late 1993.
83. Ibid., p. 241.
84. Text from EO brochure, 1993.
85. Kaplan, *Start Up*, p. 246.
86. A copy of the AT&T advertisement for the EO is available at http://www.youtube.com/watch?v=tpYhw5_LNq8, accessed 19 October 2011.
87. Henry Madden, EO product manager, email communication with the author, 17 October 2011.
88. Bradley interview, 15 February 2011.
89. Ibid.
90. Ibid.
91. Ibid.
92. Henry Madden, email communication with the author, 27 September 2011.
93. Michael Kanellos, 'Start-up Merges Cell Phone and PC into a Handheld', *CNET News*, 16 December 2005. http://news.cnet.com/Start-up-merges-cell-phone-and-PC-into-a-handheld/2100-1041_3-5997426.html, accessed 24 February 2011.
94. H. P., 'Slice of the Action', *Entrepreneur* 25/5 (May 1997), p. 22.
95. Alan Liddle, 'CyberSlice Targets Web Surfers for Pizza Delivery Business', *Nation's Restaurant News*, 16 December 1996. http://findarticles.com/p/articles/mi_m3190/is_n49_v30/ai_18976320/?tag = content;col1, accessed 10 May 2012.
96. Tim Glass, email communication with the author, 25 May 2012.
97. US Patent US7231531, 12 June 2007, filed 28 February 2003.
98. Glass email, 25 May 2012.
99. Kelly Kodama, personal communication with the author, 24 May 2012.
100. Chris Loew, email communication with the author, 23 May 2012.
101. Ibid.
102. Tim Glass, personal communication with the author, 22 May 2012.
103. Kanellos, 'Start-up Merges Cell Phone and PC'.
104. DualCor, cited in Judie Lipsett, 'Judie's Gear Diary 2006–01–05', *The Gadgeteer*. http://the-gadgeteer.com/2006/01/05/727/, accessed 10 May 2012.

105 Mike Hanlon, 'Handtop PC Combines Desktop Power, Instant-on PDA Convenience and Connected Functionality of a Cell Phone', *Gizmag* (12 January 2006). http://www.gizmag.com/go/5040/, accessed 10 May 2012.

106 CES 2006, 'Best of CES Awards: The Finalists'. http://www.cnet.com/4520–11405_1-6398234-3.html?tag=ces_index;ces_mcol, accessed 10 May 2012.

107 A video of the CES 2006 demonstration of the cPC by Tom Merritt is at http://cnettv.cnet.com/dualcor-cpc/9742-1_53-18933.html, accessed 10 May 2012.

108 Peter H. Muller, interview with the author at the studio of Interform, Woodside, CA, 11 February 2011.

109 Kanellos, 'Start-up Merges Cell Phone and PC'.

110 Glass personal communication, 22 May 2012.

MOBILE COMPUTERS

1 'In the Year 2001, the Shape of Everyday Things…', *Esquire* (May 1966), p. 116.

2 Adam Osborne, creator of the Osborne 1, quoted in M. Aartsen, 'Portable Computers, a Buyer's Guide', *Design* (March 1984), p. 48.

3 Paul Atkinson, 'Man in a Briefcase: The Social Construction of the Laptop Computer and the Emergence of a Type Form', *Journal of Design History* 18/2 (2005), pp. 191–205.

4 Rick Dickinson, interview with the author, 19 November 2010.

5 Text from a newspaper advertisement for Sinclair Executive by Sinclair, 1972.

6 The history of the various companies used by Clive Sinclair is convoluted. An off-the-shelf company bought by Sinclair was renamed Sinclair Instrument Ltd in August 1975. Sinclair asked Chris Curry to leave Sinclair Radionics and get Sinclair Instrument up and running as a fallback for when the involvement of the National Enterprise Board (who provided financial backing in return for a 43% share of Sinclair Radionics) proved too problematic. In July 1977, Sinclair Instrument was renamed Science of Cambridge Ltd. Clive Sinclair took over the running of Science of Cambridge when the National Enterprise Board closed Sinclair Radionics down and Chris Curry left to set up Acorn Computers Ltd in 1979.

7 Dickinson interview, 19 November 2010. The design of the ZX80 was started by John Pemberton and completed by Rick Dickinson. The industrial design of the ZX81 was solely Dickinson's work.

8 Barry Fox, 'Sinclair's Spectrum of Invention', *New Scientist* (25 October 1984), p. 43.

9 'Sinclair Opts for Flat Screen TV—but Cautiously', *New Scientist* (6 September 1979), p. 709.

10 'Sinclair Announces New TV—but Can He Make It?', *New Scientist* (22 September 1983), p. 856.

11 Dickinson interview, 19 November 2010.

12 'I've yet to meet anybody who thinks a liquid display is anything other than awful.' Clive Sinclair in an interview with Bill Scolding, 'Sir Clive Sinclair', *Sinclair User* 35 (February 1985). http://www.sincuser.f9.co.uk/035/sirclve.htm, accessed 10 December 2019.

13 Hugh Aldersey-Williams, 'Flat out for Pocket TV', *New Scientist* (5 May 1983), p. 282–5.

14 Dickinson interview, 19 November 2010.

15 David Redhead, *Electric Dreams: Designing for the Digital Age* (V&A Publications, 2004).

16 Martin Riddiford, interview with the author at the offices of Therefore Product Design, London, 15 December 2011.

17 Patent no. WO 98/19226, filed 28 October 1997.

18 Martin Riddiford, cited in Andrew Orlowski, 'Psion: The Last Computer. Secrets of the Sony We Never Had', *The Register*, 26 June 2007. http://www.theregister.co.uk/2007/06/26/psion_special/page7.html, accessed 8 May 2012.

19 Riddiford interview, 15 December 2011.

20 Martin Riddiford and Jim Fullalove, interview with the author at the offices of Therefore Product Design, London, 27 April 2011.

21 Ibid.

22 Originally published as Steve Litchfield, 'The History of Psion', *Palmtop Magazine* (1998); now at http://stevelitchfield.com/historyofpsion.htm, accessed 4 May 2012.

23 Ibid.

24 Riddiford and Fullalove interview, 27 April 2011.

25 Ibid.

26 Ibid.

27 Ibid.

28 Riddiford interview, 15 December 2011.

29 Riddiford and Fullalove interview, 27 April 2011.

30 Even a very recent piece in the *Daily Telegraph* newspaper used an archive image of David Potter with a Psion Series 3. See Josie Ensor, 'Psion: The Company Microsoft's Bill Gates Most Worried About', *Daily Telegraph*, 4 July 2011. http://www.telegraph.co.uk/finance/newsbysector/mediatechnologyandtelecoms/electronics/8615531/Psion-the-company-Microsofts-Bill-Gates-most-worried-about.html, accessed 4 May 2012.

31 3G (third-generation) was an emerging mobile telephone network technology at this point. 3G networks first appeared in Japan in 1998 as a non-commercial test. Although common knowledge and widely anticipated, commercial versions were not available until 2001.

32 Martin Riddiford, interview with the author at the offices of Therefore Product Design, London, 27 April 2011.

33 Jim Fullalove, interview with the author at the offices of Therefore Product Design, London, 27 April 2011.

34 Riddiford interview, 27 April 2011.

35 Ibid.

36 Fullalove interview, 27 April 2011.

37 Ibid.

38 Riddiford interview, 27 April 2011.

39 Orlowski, 'Psion'.

40 See 'The Five Coolest British Made Phones Ever—Psion Ace', *TechDigest*. http://www.techdigest.tv/2009/08/galleries/the_five_cooles.php?pic=1, accessed 4 May 2012.

41 Reuters, 'Compaq: From Place Mat Sketch to PC Giant', *USA Today*, 4 September 2001. http://www.usatoday.com/tech/techinvestor/2001-09-04-compaq-history.htm, accessed 26 April 2012.

42 Steven Stengel, 'Compaq Portable', *Obsolete Technology Website*. http://oldcomputers.net/compaqi.html, accessed 26 April 2012.

43 Reuters, 'Compaq'.

44 Morten Warren, interview with the author at the offices of Native Product Design, London, 23 January 2012.

45 Reuters, 'Compaq'.

46 Warren interview, 23 January 2012.

47 Riddiford interview, 15 December 2011.

48 Carphone Warehouse press release, 31 October 2001. http://pressoffice.carphonewarehouse.com/news/item/the_carphone_warehouse_signs_exclusive_distribution_deal_with_pogo_technolo/?phpMyAdmin=AHBslJMinbQ440hrdN48U9K1ZO9, accessed 4 February 2012.

49 Tony Smith, 'Sticking It to Pogo', *The Register*, 7 November 2001. http://www.theregister.co.uk/2001/11/07/sticking_it_to_pogo/, accessed 4 February 2012.

50 Carphone warehouse press release, 7 February 2002. http://pressoffice.carphonewarehouse.com/news/item/pogo_hits_the_high_streets/?phpMyAdmin=AHBslJMinbQ440hrdN48U9K1ZO9, accessed 4 February 2012.

51 Carphone Warehouse press release, 31 October 2001.

52 International Forum Design, ed., *iF design award 2002* (International Forum Design/Bangert Schopfheim, 2002).

53 Pogo Mobile Solutions, http://web.archive.org/web/20030402211236/http://pogomobile.com/awards/index.html, accessed 7 February 2012.

54 Kieren McCarthy, 'Pogo: Better than WAP or Just as Cwap?', *The Register*, 7 November 2001. http://www.theregister.co.uk/2001/11/07/pogo_better_than_wap/, accessed 4 February 2012.

55 Ray Le Maistre, 'Pogo Bounces Back to Life', *Light Reading Mobile*, 12 December 2002. http://www.lightreading.com/document.asp?doc_id=25747, accessed 6 February 2012.

56 Marcus Hoggarth, interview with the author at the offices of Native, London, 23 January 2012.

57   Pogo Mobile Solutions press release, 'Pogo Mobile Solutions Chooses ART's World-Leading Handwriting Recognition for nVoy', 24 June 2003. http://www.businesswire.com/news/home/20030624005342/en/Pogo-Mobile-Solutions-Chooses-ARTs-World-Leading-Handwriting, accessed 6 February 2012.

58   Access press release, 'Palm Completes Formation of Palm OS Subsidiary as Palm Powered Devices Hit 20 Million Sold'. *Access*, 21 January 2002. http://www.access-company.com/news/press/PalmSource/2002/012102.html, accessed 20 March 2012.

59   Jeff Hawkins, interview with the author at the offices of Numenta, Redwood City, CA, 14 February 2011.

60   Peter Skillman, personal communication with the author, 24 March 2012.

61   Handspring 'Hollywood', unpublished PowerPoint presentation, 20 May 2003.

62   Video available at http://allthingsd.com/video/?s = palm+foleo, accessed 21 March 2012.

63   Jeff Hawkins, quoted in Charles Arthur, 'iPhone Outsells All Other Smartphones in July, Says iSuppli', *The Guardian Technology Blog*, 5 September 2007. http://www.guardian.co.uk/technology/blog/2007/sep/05/iphoneoutsells1?INTCMP=SRCH, accessed 21 March 2012.

64   Jack Schofield, 'Palm Unveils Foleo—Reinvents Compaq Aero', *The Guardian Technology Blog*, 30 May 2007. http://www.guardian.co.uk/technology/blog/2007/may/30/palmunveilsfo, accessed 20 March 2012.

65   Todd Kort, cited in John Paczkowski, '1999 Called. It Wants Its Vadem Clio Back. Sharp Mobilon Pro, Too', *All Things D*, 4 September 2007. http://allthingsd.com/20070904/foleo-rip/, accessed 21 March 2012.

66   Tim Bajarin, 'Jeff Hawkins and the World's First Netbook', *PC World* (21 November 2008). http://www.pcmag.com/article2/0,2817,2335072,00.asp, accessed 20 March 2012.

67   Hawkins interview, 14 February 2011.

68   Ed Colligan, 'A Message to Palm Customers, Partners and Developers', *The Official Palm Blog*, 4 September 2007. http://blog.palm.com/palm/2007/09/a-message-to-pa.html, accessed 20 March 2012.

69   Skillman personal communication, 24 March 2012.

70   Hawkins interview, 14 February 2011.

71   Matt Rossof, 'HP CEO: Why We Shut Down Palm', *Business Insider*, 18 August 2011. http://articles.businessinsider.com/2011-08-18/tech/30077251_1_android-palm-hardware-hardware-business, accessed 20 March 2012.

12   Henry P. Babbage, ed., *Babbage's Calculating Engines: Being a Collection of Papers Relating to Them; Their History and Construction* (Cambridge University Press, 1889; 2010 edn).

13   Marketwire, 'The Computer History Museum Celebrates the History of Mobile Computing', *Marketwire*, 16 March 2004. http://findarticles.com/p/articles/mi_pwwi/is_20050229/ai_mark3043232181/?tag=content;col1, accessed 23 May 2012.

THE AGENCY OF IDEAS

1   Even Alan Turing's famous 1936 paper, 'On Computable Numbers', which is considered by many to contain the first description of a fully programmable computer, used the term 'computer' to refer to a human mathematician.

2   Doron Swade, 'The Construction of Charles Babbage's Difference Engine No. 2', *IEEE Annals of the History of Computing* (July–September 2005), p. 72.

3   Martin Riddiford and Jim Fullalove, interview with the author at the offices of Therefore Product Design, London, 27 April 2011.

4   Donald A. Norman, *The Invisible Computer: Why Good Products Can Fail, the Personal Computer Is So Complex and Information Appliances Are the Solution* (MIT Press, 1999), p. 40.

5   The phrase apparently comes from an old Italian saying, quoted in the film *The Desert Fox* and spoken by JFK in relation to the Bay of Pigs: 'Victory has 100 fathers but defeat is an orphan.' Its use in a business context has resulted in 'victory' and 'defeat' being replaced by 'success' and 'failure'.

6   Georgi Dalakov, 'The Difference Engine of George Grant', *History of Computers*. http://history-computer.com/Babbage/NextDifferentialEngines/Grant.html, accessed 21 May 2012.

7   See Charles Babbage, *On the Economy of Machinery and Manufactures* (Charles Knight, 1832).

8   Brian Winston, *Misunderstanding Media* (Routledge & Kegan Paul, 1986), p. 124.

9   Icon Group International, *Logarithmic: Webster's Timeline History, 1622–2007* (Icon Group International, 2009), p. 11.

10   'The Babbage Engine: Key People', *Computer History Museum*. http://www.computerhistory.org/babbage/georgedvardscheutz/, accessed 16 May 2012.

11   http://history-computer.com/Babbage/NextDifferentialEngines/Wiberg.html, accessed 16 May 2012.

Selected Bibliography

IMAGINED MACHINES

Ashley, Mike, *Out of This World: Science Fiction but Not As You Know It*, London: The British Library, 2011.

Basalla, George, *The Evolution of Technology*, Cambridge: Cambridge University Press, 1988.

Bijker, W. E., Hughes, T. P., and Pinch, T. J., eds., *The Social Construction of Technological Systems: New Directions in the Sociology and History of Technology*, Cambridge, MA: MIT Press, 1987.

Braun, Hans-Joachim, 'Symposium on Failed Innovations: Introduction', *Social Studies of Science* 22/2 (1992), pp.

Cianchi, Marco, *Leonardo da Vinci's Machines*, Florence: Becocci Editore, 1988.

Corn, Joseph J., and Horrigan, Brian, *Yesterday's Tomorrows: Past Visions of the American Future*, Baltimore: Johns Hopkins University Press, 1984.

David, P. A., 'Clio and the Economics of QWERTY', *American Economic Review* 75/2 (1985), pp.

Gelernter, David, *1939, The Lost World of the Fair*, New York: The Free Press, 1995.

Lipartito, Kenneth, 'Picturephone and the Information Age: The Social Meaning of Failure', *Technology and Culture* 44/1 (2003), pp. 51–2.

Maffei, Nicolas P., '"I Have Seen the Future": Norman Bel Geddes' "Futurama" as Immersive Design', *Design and Culture* 4/1, pp.

Pettifer, J., and Turner, N., *Automania: Man and the Motor Car*, London: Guild Publishing, 1984.

VanderMeer, Jeff, *The Steampunk Bible: An Illustrated Guide to the World of Imaginary Airships, Corsets, Goggles, Mad Scientists, and Strange Literature*, New York: Abrams Image, 2011.

Wurts, Richard, *The New York World's Fair, 1939/1940*, New York, Dover Publications, 1977.

MAINFRAMES AND MINICOMPUTERS

Baker, W. O., 'The First Ten Years of the Transistor', *The Bell System Technical Journal* 37/5 (September 1958), pp.

Carter, Frank, 'The Turing Bombe', *The Rutherford Journal* 3 (2010). http://www.rutherfordjournal.org/article030108.html, accessed 16 April 2012.

Difference Engine/Analytical Engine

Cohen, I. Bernard, 'Babbage and Aiken', *Annals of the History of Computing* 10/3 (1988), pp.

Swade, Doron, 'Automatic Computation: Charles Babbage and Computational Method', *The Rutherford Journal* 3 (2010). http://www.rutherfordjournal.org/article030106.html, accessed 9 November 2010.

Swade, Doron, 'Charles Babbage's Difference Engine No. 2 Technical Description', *Science Museum Papers in the History of Technology* 4 (1995), pp.

Swade, Doron, 'The Construction of Charles Babbage's Difference Engine No. 2', *IEEE Annals of the History of Computing* (July–September 2005), pp.

Swade, Doron, 'Redeeming Charles Babbage's Mechanical Computer', *Scientific American* (February 1993), pp.

Swade, Doron, 'The Shocking Truth about Babbage and His Calculating Engines', *Resurrection: The Bulletin of the Computer Conservation Society* 32 (2004). http://www.cs.man.ac.uk/CCS/res/res32.htm#d, accessed 9 November 2010.

Purbrick, Louise, 'The Dream Machine: Charles Babbage and His Imaginary Computers', *Journal of Design History* 6/1 (1993), pp.

Hofgaard Machine

'Hofgaardmaskinen: Et FOU-prosjekt som preget Sønnicos historie' (The Hofgaard Machine: an R&D project that characterized Sønnico's history). Sønnico Annual Report, 1986.

Nordsieck Computer

Blue, Charlie, 'A Very Modern Device', *Lawrence Livermore National Laboratory 50th Anniversary Newsletter*, 2002.

'Illini Scientist Builds "Brain" with $700', *News-Gazette*, 2 April 1950.

Nordsieck, Arnold, 'The Nordsieck Computer', *Proceedings of the Western Computer Conference—Joint IRE-AIEE-ACM* (4–6 February 1953), pp.

'U.I. Board Releases Patent on Calculator', *News-Gazette*, 26 November 1951.

Saab D2

Hallberg, Tord Jöran, *IT Gryning: Svensk datahistoria från 1840- till 1960-talet*, Lund, Sweden: Studentlitteratur, 2007.

Honeywell Kitchen Computer

Atkinson, Paul, 'The Curious Case of the Kitchen Computer: Products and Non-products in Design History', *Journal of Design History* 23/2, pp.

Marcus, Stanley, *Minding the Store*, London: Elm Tree Books, 1975.

CTL Modular Three Minicomputer

Brutton, Mark, 'Just for the Look of the Thing', *Design* 368 (August 1979).

Pawley, Martin, 'Office beneath the Skin', *Design* (March 1970).

PERSONAL COMPUTERS

Atkinson, Paul, 'Actor Networks and the Development of the Home Computer', in Fiona Hackney, Jonathan Glynne and Viv Minton, eds., *Networks of Design*, Universal Publishers, 2008.

Haddon, Leslie, 'The Home Computer: The Making of a Consumer Electronic', *Science as Culture* 2 (1988), pp.

Linzmayer, Owen, *Apple Confidential 2.0: The Definitive History of the World's Most Colorful Company*, No Starch Press, 2008.

Rushe, Dominic, 'Apple Pips Exxon as World's Biggest Company', *The Guardian*, 10 August 2011, p. 24.

IBM SCAMP Design Model

Littmann, Jonathan, 'The First Portable Computer: The Genesis of SCAMP, Grandfather of the Personal Computer', *PC World* (October 1983), p. 296 xx.

IBM Yellow Bird

Atkinson, Paul, 'Actor Networks and the Development of the Home Computer', in Fiona Hackney, Jonathan Glynne and Viv Minton, eds., *Networks of Design*, Universal Publishers, 2008.

Bush, Vannevar, 'As We May Think', *Atlantic Monthly* (July 1945).

Xerox Notetaker

'Computers: Carry Along, Punch In, Read Out', *Time* (21 June 1982). http://www.time.com/time/printout/0,8816,925484,00.html, accessed 4 December 2011.

Hiltzik, Michael, *Dealers of Lightning*, London: Orion Business, 2000.

Stobie, I., 'They All Laughed, But…', *Practical Computing* (January 1983), p. 108.

IBM 'Atari' PC

Edwards, Benj, 'Atari's Answer to the Apple II', *PC World.com* (2009). http://www.pcworld.com/article/181421/inside_the_atari_800.html, accessed 3 April 2012.

Sinclair QL+

Adamson, Ian, and Kennedy, Richard, *Sinclair and the 'Sunrise Technology'*, Penguin Books, 1986. http://www.nvg.ntnu.no/sinclair/computers/ql/ql_sst.htm, accessed 7 April 2011.

'Sinclair Exploits Old Technology for New Chip', *New Scientist* (27 September 1984), p. 27.

Dragon Professional

Coley, L., 'Dragon Professional', *Personal Computer World* (August 1984), pp.

Dragon User, May 1983, September 1983, October 1983, December 1983, August 1984, September 1984.

PEN COMPUTERS

Blickensторfer, C. H., '10 Years of Pen Computing', *Pen Computing Magazine* 50 (June 2004), pp.

Breen, B., 'Fresh Start 2002: Starting Over … and Over…', *Fast Company* 54 (December 2001), p. 77.

Fisher, M., 'Momenta Head to Offer His "Pentop" Computer', *New York Times*, 5 October 1991.

Xerox Dynabook

Hiltzik, Michael, *Dealers of Lightning*, London: Orion Business, 2000.

Kay, Alan, 'The Early History of Smalltalk', *ACM SIGPLAN Notices* 28/3 (March 1993), p. 7xxx.

Kay, Alan, 'FLEX: A Flexible Extendable Language', MSc thesis, University of Utah, 1968.

Kay, Alan, 'The Reactive Engine', PhD thesis, University of Utah, 1969.

Kay, Alan, 'A Personal Computer for Children of All Ages', *Proceedings of the ACM National Conference, Boston* (August 1972), pp.

Kay, Alan, and Goldberg, Adele, 'Personal Dynamic Media', *Computing* 10/3 (March 1977), p. 3xxxx.

Lees, John, 'The World in Your Own Notebook', in David Ahl and Burchenal Green, eds., *The Best of Creative Computing*, iii, Morristown, NJ: Creative Computing Press, 1980.

Apple Figaro

Dair, Tom, 'Smart Design's iPad … Circa 1989', *Fast Company* (26 March 2010). http://www.fastcompany.com/1598501/smart-designs-ipad-circa-1989, accessed 24 April 2012.

Kunkel, Paul, *Apple Design: The Work of the Apple Industrial Design Group*, New York: Graphis US Inc., 1997.

Sun Modular Computer

Arthur, Charles, 'Jonathan Schwartz Tweets His Last Goodbye to Sun Microsystems', *The Guardian Technology Blog*, 4 February 2010. http://www.guardian.co.uk/technology/blog/2010/feb/04/jonathan-schwartz-sun-microsystems-tweet-ceo-resignation, accessed 12 March 2012.

Ghosh, Rishab Aiyer, *Study on the Economic Impact of Open Source Software on Innovation and the Competitiveness of the Information and Communication Technologies (ICT) Sector in the EU*, European Commission report 0.5754, 2006. http://ec.europa.eu, accessed 16 March 2012.

Schofield, Jack, 'Larry Ellison Starts to Reign over Sun', *The Guardian Technology Blog*, 13 May 2010. http://www.guardian.co.uk/technology/blog/2010/may/13/ellison-oracle-sun/print, accessed 12 March 2012

GO PenPoint Computer/EO Magni

Edstrom, Jennifer, and Eller, Marlin, *Barbarians Led by Bill Gates: Microsoft from the Inside*, New York: H. Holt, 1998.

Goodwins, Rupert, 'Go PenPoint', *Personal Computer World* (April 1991), p. 140.

Kaplan, Jerry, *Start Up: A Silicon Valley Adventure*, London: Little, Brown and Co., 1994.

IBM Leapfrog Tablet

Hardy, Tom, 'Innovation and Chaos', *Design Management Journal* 5/3 (Summer 1994), pp.

Romanelli, Marco, 'Leapfrog Computer', *Domus* (February 1994), pp.

Sakakibara, Kiyonori, 'Global New Product Development: The Case of IBM Notebook Computers', *Business Strategy Review* 6/2 (1995), pp.

Trelstad, Julie, 'Code Name Leapfrog', *I.D.: The International Design Magazine* (May–June 1993), pp.

DualCor cPC

Kanellos, Michael, 'Start-up Merges Cell Phone and PC into a Handheld', *CNET News*, 16 December 2005. http://news.cnet.com/Start-up-merges-cell-phone-and-PC-into-a-handheld/2100-1041_3-5997426.html, accessed 24 February 2011.

MOBILE COMPUTERS

Aartsen, M., 'Portable Computers, A Buyer's Guide', *Design* (March 1984), pp.

Atkinson, Paul, 'Man in a Briefcase: The Social Construction of the Laptop Computer and the Emergence of a Type Form', *Journal of Design History* 18/2 (2005), pp. 191–205.

'In the Year 2001, the Shape of Everyday Things…', *Esquire* (May 1966), p. 116.

Sinclair Pandora Laptop

Aldersey-Williams, Hugh, 'Flat out for Pocket TV', *New Scientist* (5 May 1983), p. 282–5.

Fox, Barry, 'Sinclair's Spectrum of Invention', *New Scientist* (25 October 1984), p. 43.

Scolding, Bill, 'Sir Clive Sinclair', *Sinclair User* 35 (February 1985). http://www.sincuser.f9.co.uk/035/sirclve.htm, accessed 10 December 2010.

'Sinclair Announces New TV—but Can He Make It?', *New Scientist* (22 September 1983), p. 856.

'Sinclair Opts for Flat Screen TV—but Cautiously', *New Scientist* (6 September 1979), p. 709.

Phonebook/Psion Halo and Ace

Orlowski, Andrew, 'Psion: The Last Computer: Secrets of the Sony We Never Had', *The Register*, 26 June 2007. http://www.theregister.co.uk/2007/06/26/psion_special/page7.html, accessed 8 May 2012.

Redhead, David, *Electric Dreams: Designing for the Digital Age*, London: V&A Publications, 2004.

Siemens PIC

Litchfield, Steve, 'The History of Psion', *Palmtop Magazine* (1998).

Compaq Dualworlds Notebook

Reuters, 'Compaq: From Place Mat Sketch to PC Giant', *USA Today*, 4 September 2001. http://www.usatoday.com/tech/techinvestor/2001-09-04-compaq-history.htm, accessed 26 April 2012.

Pogo nVoy Communicator

Le Maistre, Ray, 'Pogo Bounces Back to Life', *Light Reading Mobile*, 12 December 2002. http://www.lightreading.com/document.asp?doc_id=25747, accessed 6 February 2012.

McCarthy, Kieren, 'Pogo: Better than WAP or Just as Cwap?', *The Register*, 7 November 2001. http://www.theregister.co.uk/2001/11/07/pogo_better_than_wap/, accessed 4 February 2012.

Smith, Tony, 'Sticking It to Pogo', *The Register*, 7 November 2001. http://www.theregister.co.uk/2001/11/07/sticking_it_to_pogo/, accessed 4 February 2012.

Palm Foleo

Arthur, Charles, 'iPhone Outsells All Other Smartphones in July, Says iSuppli', *The Guardian Technology Blog*, 5 September 2007. http://www.guardian.co.uk/technology/blog/2007/sep/05/iphoneoutsells1?INTCMP=SRCH, accessed 21 March 2012.

Bajarin, Tim, 'Jeff Hawkins and the World's First Netbook', *PC World* (21 November 2008). http://www.pcmag.com/article2/0,2817,2335072,00.asp, accessed 20 March 2012.

Paczkowski, John, '1999 Called. It Wants Its Vadem Clio Back. Sharp Mobilon Pro, Too', *All Things D*, 4 September 2007. http://allthingsd.com/20070904/foleo-rip/, accessed 21 March 2012.

Rossof, Matt, 'HP CEO: Why We Shut Down Palm', *Business Insider*, 18 August 2011. http://articles.businessinsider.com/2011-08-18/tech/30077251_1_android-palm-hardware-hardware-business, accessed 20 March 2012.

Schofield, Jack, 'Palm Unveils Foleo—Reinvents Compaq Aero', *The Guardian Technology Blog*, 30 May 2007. http://www.guardian.co.uk/technology/blog/2007/may/30/palmunveilsfo, accessed 20 March 2012.

THE AGENCY OF IDEAS

Babbage, Charles, *On the Economy of Machinery and Manufactures*, London: Charles Knight, 1832.

Babbage, Henry P., ed., *Babbage's Calculating Engines: Being a Collection of Papers Relating to Them; Their History and Construction*, Cambridge: Cambridge University Press, 1889 (2010 edn).

Icon Group International, *Logarithmic: Webster's Timeline History, 1622–2007*, Icon Group International, 2009.

Marketwire, 'The Computer History Museum Celebrates the History of Mobile Computing', *Marketwire*, 16 March 2004.

Norman, Donald A., *The Invisible Computer: Why Good Products Can Fail, the Personal Computer Is So Complex and Information Appliances Are the Solution*, Cambridge, MA: MIT Press, 1999.

Winston, Brian, *Misunderstanding Media*, London: Routledge & Kegan Paul, 1986.

Index

Advanced Research Projects Agency (ARPA), 106, 110
Aiken, Howard, 25
Airy, George Biddell, 16–17
Altair, see MITS
Amstrad, 97, 167
Analytical Engine, see Babbage, Charles
Apple Computers Inc, 79, 107, 114–23, 135, 145, 221–2
    Apple Figaro, 114–23, 213, 222
    Apple iMac, 222
    Apple iPad, 107, 109, 210, 212, 223
    Apple iPhone, 107, 159, 223
    Apple Lisa, 83, 104, 221
    Apple Macintosh, 83, 95, 221
    Apple Newton/Newton MessagePad, 107, 135, 145, 149, 213, 222
    Apple 1, 60, 70, 89, 220
    Apple III, 221
    Apple][(Apple II), 60–1, 76, 78, 88, 102, 220–1
Applied Computer Techniques (Apricot Computers), 59
Ashley, Mike, 8
Atanasoff-Berry Computer (ABC), 218
AT&T, 9, 126, 135, 143, 145, 147–9
Atari, 60, 87–91
    Atari 400, 89, 221
    Atari 800, 88–9, 91, 221
Augmentation Research Center, 113

Babbage, Charles, 14–25, 210–15
    Analytical Engine, 17, 20–5, 210, 218
    Difference Engine, 14–19, 21, 23–5, 210, 213, 215, 218
Babbage, Henry Prevost, 25, 215
Backs, Jochen, 124–6
Bacon, Roger, 2
Baranski, Celeste, 132, 145
BARK, 38
Barron, Iann, 55, 57, 59
Bassalla, George, 4
Bechtolsheim, Andreas, 126
Bel Geddes, Norman, 5–6
Bell, Gordon, 50, 51, 215
Bell Laboratories, 9, 33, 77, 218
Berkeley Computer Corporation, 113
Berners-Lee, Tim, 221
BESK, 38–9
Bijker, Wiebe, 10
Bitzer, Donald, 110
Blackberry, 159, 174
Bletchley Park, 12, 218
Bloch, Felix, 33
Bockler, George Andreas, 3
Booker, Sue, 116–18, 123
Bradley, Paul, 124–6, 130–2, 142, 147–9
Braun, Hans-Joachim, 10

Brett, Graham, 170
British Broadcasting Corporation (BBC), 187
    BBC Micro, 104, 221
Brunner, Robert, 118, 123
Bushnell, Nolan, 88–9
Byte, 215

Cambridge Computers, see Sinclair, Clive
Canion, Rod, 190
Cappelen, August, 29
Carphone Warehouse, 198, 203
Carr, Robert, 132
Catt, Ivor, 97
Cheadle, Ed, 110
Clarke, Arthur, C., 11
Clarke, Tony, 102, 104
Clements, John, 17, 213
CNET, 153
Colligan, Ed, 206, 209
Colossus, 12, 218
Commodore Business Machines/Commodore International Ltd., 79, 102
    Commodore PET 2001, 60, 76, 88, 220
    Commodore 64, 221
    Commodore SX-64, 215, 217
Compaq Computer Corporation, 131, 188–95
    Compaq Aero 8000, 209
    Compaq Deskpro, 191
    Compaq Dualworlds, 188–95, 211, 222
    Compaq Portable, 190–1, 215, 217, 221
Computer Controls Corporation (CCC), 50
Computer History Museum, 34, 51
Computer Space, 88
Computer Technology Limited (CTL), 52, 55, 57, 59
    CTL Modular One, 53–5, 57
    CTL Modular Three, 52–9, 220
    CTL Modular Two, 55, 57
Cray-1, 220
Cupps, Bryan, 152–3
Cyberslice Inc., 152

Dair, Tom, 118
Danielsen, George, 29
DASK, 39
Data General, 50–1, 55
    Data General Nova Minicomputer, 50, 220
Datasaab, 43
David, Paul, A., 10
David Kelly Design, 132
Da Vinci, Leonardo, 2–4
Dell Computers, 195
Design Council, 163, 198
Dickinson, Rick, 92, 95–7, 160, 163–4

Difference Engine, see Babbage, Charles
Difference Engine, The, 212
Digital Equipment Corporation (DEC), 51, 55, 60, 215
    DEC PDP1, 219
Doren, Kevin, 132
Dragon Data Ltd., 100–5
    Dragon Professional/GEC Dragon Professional, 100–5, 221
    Dragon 64/GEC Dragon 64, 103–5
    Dragon 32, 101–5
    Dragon User, 103, 105
DualCor Technologies, 150–7, 211
    DualCor cPC, 150–7, 211, 223
Dubinsky, Donna, 206

Eames, Charles, 138
Earl, Harley, 6
Edison, Thomas, 9
EDSAC, 218
Ellenby, John, 158
Elliott, John, 52, 57–8
Elliott-Automation, 55
Engelbart, Doug, 110, 113, 219–20
ENIAC, 12, 33, 218
EO Corporation, 135, 142–9
    EO 880, 145, 147
    EO 440, 143–9, 222
    EO Loki, 149
    EO Magni Personal Communicator, 135, 142–9, 211, 222
Ericsson, 43, 174, 179, 184
Esslinger, Hartmut, 116–17, 145
Eurohard SA, 105
Everyday Science and Mechanics, 7

Facit EDB, 39–41
Fairbairn, Doug, 83
Felsenstein, Lee, 215
Ferranti Mark 1, 12, 218
FLEX, 110, 113
Frazer Designers, 170
Friedl, Paul J., 64–6
Frigidaire, 6
Fukusawa, Naoto, 142, 147–9
Fullalove, Jim, 180
Fuller, Buckminster, 6
Function Engineering, 153

Gabor, Dennis, 164
Gadgeteer, The, 153
Gardner, Hendrie, 50
Garvis, Jerry, 66
Gassée, Jean-Louis, 118
Gates, Bill, 132, 135

General Electric Company (GEC), 100–5
General Motors, 5–6, 8, 34
    Futurama, 5
    Motorama, 6
George, Joe, 65
Gernsback, Hugo, 9
Giugiaro, Giorgietto/ Giugiaro Design, 114–23
Gizmag, 155
Glass, Tim, 152, 157
GO Corporation, 130–5, 145–9, 215–16
    GO Penpoint Computer, 130–5, 145, 211, 215–16, 221
    PenPoint, 132–5, 145
Goldberg, Adele, 83
GRAIL, 106–10
Grant, George Bernard, 215
GRiD Computers, 206
    GRiD Compass, 82–3, 158, 221
    GRiDPad, 106–7, 206, 221

Haitani, Rob, 206
Hammel, Paul, 145
Handspring, *see* Palm Computing
Hanley, Steve, 153–5
Hardy, Tom, 62–3, 66–8, 70, 73–4, 76–9, 86–8, 90–1, 138, 140
Harris, Jim, 190
Hawkins, Jeff, 107, 204, 206–9, 211
Heisenberg, Werner, 33
Herschel, John, 15
Hewlett-Packard Company (HP), 60, 134, 181, 189, 195, 209
Hofgaard, Rolf, 26–9, 211
    Hofgaard Machine, The, 26–9, 210–11, 218
Hoggarth, Marcus, 196, 198, 203
Hollerith Tabulating Machines, 12, 23
Homebrew Computer Club, 60, 220
Honeywell Computer Control Division, 44–6, 50–1, 158, 215
    Honeywell H316, 46–8, 50–1
    Honeywell Kitchen Computer, 44–51, 213, 215, 220

IBM, 13, 25, 27, 38, 60–79, 86–91, 104, 126, 129, 134–40, 145, 181, 190–1, 210–13, 215, 221
    IBM Advanced Systems Development Laboratory, 65
    IBM Aquarius, 74–9, 211, 220
    IBM ASCC (Harvard Mk1), 12, 24–5, 218
    IBM Atari PC, 86–91, 213, 221
    IBM Entry Level Systems, 66, 88
    IBM 5100, 66–7, 70, 213
    IBM 5110, 67, 220
    IBM 5150 (IBM PC), 61, 67, 87, 91, 95, 104, 126, 190, 213, 215, 221
    IBM General Systems Division (GSD), 63–4, 70, 76, 88
    IBM Leapfrog, 136–41, 221
    IBM PALM, 65
    IBM Palo Alto Scientific Center, 64
    IBM SCAMP, 64, 70, 83, 88, 213
    IBM SCAMP Design Model, 62–5, 220
    IBM 701, 219
    IBM SSEC, 218
    IBM System/360, 13, 43, 219
    IBM ThinkPad, 138, 140
    IBM Yellow Bird, 68–73, 78, 220
Ideal Home Exhibition, 6
Intel, 59, 132, 135, 220
    Pentium processor, 222
Interform, 156
International Computers Ltd. (ICL), 95
    ICL One Per Desk (ICL OPD), 95
iWare, 152

Jobs, Steve, 60, 70, 83, 152, 221–2
Jordan, Richard, 117–18
Joy, Bill, 126

Kaplan, Jerry, 131–5
Kapor, Mitchell, 131–2
Kay, Alan, 80–5, 108–13, 190, 211, 215–16
Kaypro II, 215–16
Kelemen, Don, 44, 47, 50–1
Kenbak-1, 220
Khosla, Vinod, 126
Kilby, Jack, 219
Kodama, Kelly, 152–3
Kort, Todd, 209
Kubrick, Stanley, 9, 158
    2001: A Space Odyssey, 9, 50, 158

Lampson, Butler, 113
Learning Research Group, 113
Linus Technologies, 106
    Linus Write-Top, 106, 221
Linux, 206, 222
Lipartito, Kenneth, 10
Loew, Chris, 153–4
LOGO, 110
Lotus Development Corporation, 131
Lowe, Bill, 64, 66, 70, 73, 76–9, 88–9, 91
Lucente, Sam, 136, 138
Lyon's Electronic Office (LEO), 218

McKinsey, Kevin, 88–9
McNealy, Scott, 126
Madden, Henry, 149
Magnavox Odyssey, 88
Manchester 'Baby', 218
Manchester Mark 1, 218
Matrix Product Design, 124–32
Melchor, Jack, 215

Mettoy Ltd., 102–4
Microsoft, 107, 132–5, 175, 181, 190, 203, 213, 216, 221
    Microsoft Windows, 132, 135, 153
        Windows CE, 177, 180–1, 211
        Windows Mobile 5.0, 153
        Windows 3, 221
        Windows XP, 153
Minsky, Marvin, 110
MITS, 60, 220
    Altair 8800, 60, 70, 220
Modern Mechanix, 6
Moggridge, Bill, 52–3, 55, 57, 158
    Moggridge Associates, 52–3, 55
Mokady, Ran, 203
Momenta, 106–7
Moore, Brian, 104
Moore, Gordon, 60, 219
    Moore's Law, 110, 219
Motorola, 59, 173–4, 179, 186
Muller, Peter H., 156–7
Murto, Bill, 190

National Cash Registers (NCR), 145
National Enterprise Board, 59, 164
National Research Development Corporation, 164
Negroponte, Nicholas, 113
Neiman Marcus, 45, 49–51, 215
Netscape, 222
New Scientist, 164
NeXT, 152, 221
Nokia, 172, 179, 184, 186
Nordsieck, Arnold, 30, 33–4
    Nordsieck Computer, The, 30–5, 211, 218
Noyes, Eliot, 138

Olsen, Ken, 60, 215
Opel, John, 66
Oppegaard, K. F., 28–9
Oppenheimer, J. Robert, 33
Oracle Corporation, 129
Osborne, Adam, 215–16
    Osbourne Computer Corporation, 215–16
    Osbourne 1, 83, 85, 190, 215–16, 221
Ouye, Mike, 132, 145

Palm, Conny, 38
Palm Computing, 107, 204–9, 211, 222
    Handspring, 206
    Palm Foleo, 204–9, 211, 223
    PalmOne, 206
    Palm Pilot, 107, 206, 222
    PalmSource, 206
    Palm Treo, 206

Papert, Seymour, 110, 113
PA Technology (PAT), 102
Path Dependency, 10
Patton, Doug, 118
Pen Computing Magazine, 107
PenPoint, *see* GO Corporation
Personal Computer World, 104–5, 134
Philco-Ford, 6
Philips, 105, 174
Phonebook, 168–75, 211, 213, 222
Pilot ACE, 218
Pinch, Trevor, 10
Pogo Technology Ltd., 198–203
    Pogo Mobile Solutions, 196, 203
    Pogo nVoy Communicator, 196–203, 211, 222
Pong, 88, 220
Popular Electronics, 70
Potter, David, 170, 178–9, 184
Psion Computers, 170–3, 178–87, 211, 221
    Psion Halo and Ace, 182–7, 213, 222
    Psion Organiser, 171–2, 178, 221
    Psion Series 5, 178
    Psion Series 7, 186, 209
    Psion Series 3/3a, 170, 178, 184

RAND, 106, 110
Rand, Paul, 138
Razorfish, 198
Register, The, 169
Riddiford, Martin, 168, 172, 176, 178, 180, 182, 184–5
Rosen, Ben, 190

Saab, 36–43
    Saab CK37, 42–3
    Saab D2, 36–7, 41, 43, 213, 219
    Saab D21, D22 and D23, 42–3
Saab-Univac, 43
SAGE, 106
Sakoman, Steve, 118
SANK, 40, 43
Sapper, Richard, 136–8
SARA, 38, 40–1
Scheutz, Pehr Georg, 213
    Scheutzian Calculation Engine, 213–14
Schwartz, Jonathan, 129
Science and Invention, 6
Science Wonder Stories, 9
Sculley, John, 115–16, 118, 123, 145
Shirai, Greg, 206
Siemens AG, 106, 174, 176–7, 179–81
    Siemens PIC, 176–81, 211, 222
Sinclair, Clive, 93–7, 162–4, 167
    Anamartic Ltd., 97, 213

Cambridge Computers Z88, 167, 213
Science of Cambridge, 163
Sinclair Computers, 163
Sinclair Executive, 162
Sinclair Flat Screen TV/TV80, 97, 165
Sinclair Mega PC Waferstack, 97, 99
Sinclair Microdrive, 95, 97, 161, 163–4, 167
Sinclair Microvision, 162–4
Sinclair MK14, 163
Sinclair Pandora Laptop, 97, 160–7, 211, 213, 221
Sinclair QL, 94–7, 105, 178, 221
Sinclair QL+, 92–7, 221
Sinclair Radionics, 162, 164
Sinclair Research Ltd., 92–9, 102, 160–7, 178, 211, 213, 221
Sinclair Super QL, 97–8
Sinclair ZX80, 95, 162–3, 221
Sinclair ZX81, 95, 163, 178, 221
Sinclair ZX Spectrum/ Spectrum+, Spectrum 128, 95, 102, 161, 163, 178, 221
Sinclair, Ian, 162
Sketchpad, 106, 110, 219
Skillman, Peter, 204, 206, 209
Smalltalk, 85, 111, 113
Smart Design, 116–18
Social Construction of Technology, 9–10, 216
Sønnichsen & Co./Sønnico, 26–9, 210
    Sønnichsen, Francis, 28
Sottsass, Ettore, 117–18
Spicer, Dag, 6
State Farm Mutual Automobile Insurance, 134–5
Steampunk Manifesto, 11
Sugar, Alan, 167
SUN Microsystems, 124–9, 152
    SUN Modular Computer, 124–9, 221
Sutherland, Ivan, 110, 219
Swade, Doron, 15, 17
Symbian Ltd., 179–80, 184, 186
Syzygy Engineering, 88

Talerico, Joe, 66
Tandy Corporation, 76, 79, 105
    Tandy TRS-80, 60, 76, 88, 102
Tchao, Michael, 123
Technological Determinism, 9–10
Teklogix, 187
Tesla, Nikola, 9
Tesler, Larry, 83, 85
Texas Instruments, 59, 190, 219
Thacker, Chuck, 113
Therefore Product Design Consultants, 168–203, 211, 213
Thorne, Wink, 135
Thorn-EMI, 59
3Com, 206

T-Mobile, 172, 174–5
Tomlinson, Ray, 220
Torvalds, Linus, 222
Turing, Alan, 12, 16, 218
TX-0, 219
TX-2, 106, 219

UNIVAC, 12, 218
UNIX, 126, 220
US Robotics, 206

Verne, Jules Gabriel, 3–4
Videophone, 9
VisiCalc, 61, 88, 221
Vodafone, 172, 175
Von Neumann, John, 23, 38

Wang, 134
Warner Communications, 88
Warren, Morten, 188–91, 195
Watson, Thomas, 60
Watson, Thomas, Jr., 138
Whirlwind, 106
Wiberg, Martin, 213
Williamson, Ian, 163
Windows, *see* Microsoft
Witts, Martin, 176, 180
Wood, Ken, 123
World Wide Web, 221
Wozniac, Steve, 60

Xerox Corporation, 80–5, 108–13, 158, 211–12, 215–16, 220
    Xerox Alto, 84, 113, 220
    Xerox Dynabook, 83, 108–13, 220
    Xerox Notetaker, 80–5, 113, 190, 215–16, 220
    Xerox Palo Alto Research Center (PARC), 83, 85, 113, 215
    Xerox Star, 221

Yahoo, 222

Zoe Design Associates, 153
Zuse, Konrad, 218